CONNECTIONS

The EERI Oral History Series

LeRoy Crandall

CONNECTIONS

The EERI Oral History Series

LeRoy Crandall

Stanley Scott, Interviewer

 Earthquake Engineering Research Institute

Editor: Gail Hynes Shea, Berkeley, California, www.gailshea.com

Cover and book design: Laura H. Moger, Moorpark, California, www.lauramoger.com

Copyright 2008 by the Earthquake Engineering Research Institute

The publication of this book was supported by FEMA/U.S. Department of Homeland Security under grant #EMW-2004-CA-0297.

The opinions expressed in this publication are those of the oral history subject and do not necessarily reflect the opinions or policies of the Earthquake Engineering Research Institute or the University of California.

Published by the Earthquake Engineering Research Institute

> 499 14th Street, Suite 320
> Oakland, CA 94612-1934
> Tel: (510) 451-0905   Fax: (510) 451-5411
> E-mail: eeri@eeri.org
> Website: http://www.eeri.org

EERI Publication Number: OHS-15

Library of Congress Cataloging-in-Publication Data
Crandall, LeRoy, 1917-
  LeRoy Crandall / Stanley Scott, interviewer.
      p. cm. -- (Connections: the EERI oral history series ; 15)
  Includes index.
  "This oral history volume is the completion of the interview sessions Stanley Scott (1921-2002) conducted with LeRoy Crandall in 1989, 1990, and 1991, which provide most of the content of this book, and in a final interview between Scott and Crandall in 2000 that is included as the last chapter of this volume"--Forward.
  ISBN 978-1-932884-38-8 (pbk. : alk. paper)
  1. Crandall, LeRoy, 1917---Interviews. 2. Civil engineers--California--Interviews.
3. Earthquake engineering--California--History. I. Scott, Stanley, 1921-2002. II. Earthquake Engineering Research Institute. III. Title.
  TA140.C74C73 2008
  624.1'51092--dc22
  [B]
                                2008034716
Printed in the United States of America

1 2 3 4 5 6 7 8 20 19 18 17 16 15 14 13 12 11 10 09 08

Table of Contents

The EERI Oral History Series

This is the fifteenth volume in the Earthquake Engineering Research Institute's series, *Connections: The EERI Oral History Series*. EERI began this series to preserve the recollections of some of those who have had pioneering careers in the field of earthquake engineering. Significant, even revolutionary, changes have occurred in earthquake engineering since individuals first began thinking in modern, scientific ways about how to protect construction and society from earthquakes. The *Connections* series helps document this important history.

Connections is a vehicle for transmitting the fascinating accounts of individuals who were present at the beginning of important developments in the field, documenting sometimes little-known facts about this history, and recording their impressions, judgments, and experiences from a personal standpoint. These reminiscences are themselves a vital contribution to our understanding of where our current state of knowledge came from and how the overall goal of reducing earthquake losses has been advanced. The Earthquake Engineering Research Institute, incorporated in 1948 as a nonprofit organization to provide an institutional base for the then-young field of earthquake engineering, is proud to help tell the story of the development of earthquake engineering through the *Connections* series. EERI has grown from a few dozen individuals in a field that lacked any significant research funding to an organization with nearly 3,000 members. It is still devoted to its original goal of investigating the effects of destructive earthquakes and publishing the results through its reconnaissance report series. EERI brings researchers and practitioners together to exchange information at its annual meetings and, via a now-extensive calendar of conferences and workshops, provides a forum through which individuals and organizations of various disciplinary backgrounds can work together for increased seismic safety.

The EERI oral history program was initiated by Stanley Scott (1921-2002). The first nine volumes were published during his lifetime, and manuscripts and interview transcripts he left to EERI are resulting in the publication of other volumes for which he is being posthumously credited. In addition, the Oral

History Committee is including further interviewees within the program's scope, following the Committee's charge to include subjects who: 1) have made an outstanding career-long contribution to earthquake engineering, 2) have valuable first-person accounts to offer concerning the history of earthquake engineering, and 3) whose backgrounds, considering the series as a whole, appropriately span the various disciplines that are included in the field of earthquake engineering.

Scott's work, which he began in 1984, summed to hundreds of hours of taped interview sessions and thousands of pages of transcripts. Were it not for him, valuable facts and recollections would already have been lost.

Scott was a research political scientist at the Institute of Governmental Studies at the University of California at Berkeley. He was active in developing seismic safety policy for many years, and was a member of the California Seismic Safety Commission from 1975 to 1993. Partly for that work, he received the Alfred E. Alquist Award from the Earthquake Safety Foundation in 1990.

Scott received assistance in formulating his oral history plans from Willa Baum, Director of the University of California at Berkeley Regional Oral History Office, a division of the Bancroft Library. Following his retirement from the University in 1989, Scott continued the oral history project. For a time, some expenses were paid from a small grant from the National Science Foundation, but Scott did most of the work pro bono. This work included not only the obvious effort of preparing for and conducting the interviews themselves, but also the more time-consuming tasks of reviewing transcripts and editing the manuscripts to flow smoothly.

The *Connections* oral history series presents a selection of senior individuals in earthquake engineering who were present at the beginning of the modern era of the field. The term "earthquake engineering" as used here has the same meaning as in the name of EERI—the broadly construed set of disciplines, including geosciences and social sciences as well as engineering itself, that together form a related body of knowledge and collection of individuals that revolve around the subject of earthquakes. The events described in these oral histories span many kinds of activities: research, design projects, public policy, broad social aspects, and education, as well as interesting personal aspects of the subjects' lives.

Published volumes in *Connections: The EERI Oral History Series*

Henry J. Degenkolb	1994
John A. Blume	1994
Michael V. Pregnoff and John E. Rinne	1996
George W. Housner	1997
William W. Moore	1998
Robert E. Wallace	1999
Nicholas F. Forell	2000
Henry J. Brunnier and Charles De Maria	2001
Egor P. Popov	2001
Clarence R. Allen	2002
Joseph Penzien	2004
Robert Park and Thomas Paulay	2006
Clarkson W. Pinkham	2006
Joseph P. Nicoletti	2006
LeRoy Crandall	2008

EERI Oral History Committee

Robert Reitherman, Chair
William Anderson
Roger Borcherdt
Gregg Brandow
Ricardo Dobry
Robert Hanson
Loring A. Wyllie, Jr.

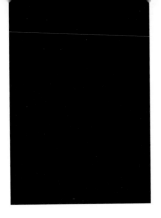

Foreword

This oral history volume is the culmination of interview sessions Stanley Scott (1921-2002) conducted with LeRoy Crandall in 1989, 1990, and 1991, which provide most of the content of this book, and in a final interview between Scott and Crandall in 2000 that is included as the last chapter in this volume. I edited and reorganized the manuscript to condense and place together discussions that occurred at different times and are related to the same topic. That editing did not change the substance of what was said, and in cases where it is important to know the date when the interview occurred, it is noted. Footnotes and photographs have also been added to complete the work. Two members of the Oral History Committee, Loring Wyllie and Ricardo Dobry, reviewed the manuscript. In addition to writing the personal introduction, Marshall Lew also reviewed a draft and provided comments and corrections.

Gail Shea, consulting editor to EERI, carefully reviewed the entire manuscript and prepared the index, as she has on previous *Connections* volumes, and Eloise Gilland, the Editorial and Publications Manager of EERI, also assisted in seeing this publication through to completion.

Robert Reitherman
Chair, EERI Oral History Committee
June 2008

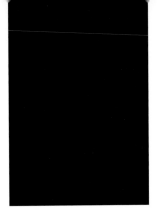

Personal Introduction

I have had the privilege to know LeRoy Crandall for almost all of my professional career as a geotechnical and earthquake engineer. I interned at LeRoy Crandall and Associates in the early 1970s while a graduate student at UCLA, and later joined his firm on a full-time basis in 1977 after one year as an underpaid Assistant Professor at California State University, Long Beach.

LeRoy was the engineer's engineer. He knew everyone and everyone knew him, or at least knew of him. He was connected with every big name architect and structural engineer in southern California. He was also known as the King of Downtown Los Angeles and Mr. High-rise, as LeRoy was the geotechnical engineer for almost every high-rise building in downtown Los Angeles and other areas in southern California during the heyday of tall buildings in the 1960s, 70s, 80s, and into the 90s.

LeRoy was a great person to work for. He surrounded himself with talented associates that formed the nucleus of what was the most prominent geotechnical consulting firm in southern California for decades, LeRoy Crandall and Associates. His earliest associates were Fred Barnes, Leopold Hirschfeldt, and Russ Weber; together they were the "Big Four." They were later joined by Jimmy Kirkgard, Jim McWee, Perry Maljian, Seymour Chiu, Robert Chieruzzi, and Jim van Beveren. Glenn Brown joined as an associate when LeRoy Crandall and Associates merged with Glenn A. Brown and Associates to add engineering geology expertise to the company. I was the last associate to join in 1979. LeRoy Crandall and Associates was supported by the most loyal employees, who worked long and hard to provide the best service to LeRoy's clients. Working for LeRoy was like working for family, and he treated everyone with respect and concern for their well-being.

LeRoy was and still is hard working—since his retirement from Law/Crandall in 1999, LeRoy has enjoyed his "retirement" by not working Saturdays and Sundays. His business ethic is "Do the work right and don't worry about the budget," because in the end, LeRoy believes that everything will work out.

LeRoy was not afraid to innovate. He pioneered the use of tied-back shoring in southern California, which made possible very deep excavations for the new high-rise buildings when Los Angeles eliminated the 13-story height limit in the 1950s. LeRoy was also involved with the planning and development of the first base-isolated building in the United States, the Foothill Communities Law and Justice Center in Rancho Cucamonga, San Bernardino County, California, not far from the San Andreas fault.

He emphasized professionalism and instilled a sense of pride in our work. He also encouraged participation in professional societies and giving to the community. He served on the Board of the Los Angeles YMCA and had a remarkable attendance record for his fifty years in Rotary International. He was heavily involved with the Structural Engineers Association of Southern California and served on the Board of Directors of the Earthquake Engineering Research Institute, American Society of Civil Engineers, and American Council of Engineering Companies. LeRoy encouraged his associates and employees to also serve in professional societies and contribute to the profession and the public. He was instrumental in the establishment of the ASCE Technical Council on Lifeline Earthquake Engineering. He was an early supporter of the California Strong Motion Instrumentation Program and was appointed to the California Seismic Safety Commission by Governor Jerry Brown and reappointed by Governor Ronald Reagan. His contributions to seismic safety and geotechnical engineering are generally unheralded, but are significant and visionary.

It is hard to not love LeRoy Crandall. He is not afraid to kick you in the rear end when you need it, but he is always encouraging and compassionate. His enthusiasm is contagious, and he is an inspiration.

Marshall Lew
MACTEC Engineering and Consulting, Inc.
Los Angeles, California
November 2007

CONNECTIONS

The EERI Oral History Series

LeRoy Crandall

Chapter 1

Early Years Through High School

I am Lionel LeRoy Crandall, and with that name I often thought my parents must have anticipated I would be poet laureate of the United States.

Crandall: I am Lionel LeRoy Crandall, and with that name I often thought my parents must have anticipated I would be poet laureate of the United States. I was born on February 4, 1917, in Portland, Oregon. I have a brother two and a half years older than I, named Clifford. Unfortunately, my mother passed away shortly after I was born. The family moved to San Diego when I was a few months old, so my entire childhood was spent in San Diego. I still think it is one of the greatest places where one can grow up.

My father married again, and my brother stayed with him. I was raised by my paternal grandparents. Later on, when I was six, my brother also came to live with us. So my grandparents raised both my brother and me.

My early recollections were going to grammar school in San Diego. I attended Jefferson Grammar School, then

Roosevelt Junior High School and San Diego High School. I was a good student and enjoyed school. I did better in grammar school and junior high, particularly in junior high, where I was one of the top honor students, and I was president of my seventh grade class. I was heavily into activities with the dramatic club.

Then I went to San Diego High. Other activities seemed to enter into my life, and I didn't put as much time into the scholarly work.

Scott: You spent more time on your social life?

Crandall: No, not really. I was in the dramatic club and was in some plays, but the family wasn't wealthy, so I worked. I had a paper route, and did things of that sort. On Saturdays I worked in Safeway stores—called Heller Stores in those days in San Diego—doing things like sacking potatoes for a dollar a day.

I was also a pal of my older brother and others in his age group. They weren't particularly interested in school—especially my brother. He graduated from high school, but wasn't out to set any records scholastically.

So in short, I just didn't devote the time to school. But I did get out of high school with a B+ average. I just did not have all As, as I had in junior high. I had a couple of student body positions in high school, but mostly I was in the dramatic club.

As far as influences on me in school, two teachers in San Diego High School were the most important. A Miss Cupp was the English teacher. She was a hard taskmaster, but we really learned a great deal about English in that class. The other was a physics teacher, Rex

Doughty, whom I admired. We got along fine, partly because his name was Rex, which means "king" in Latin, and my name, LeRoy, also means king, in French. I was very interested in Latin class. I enjoyed physics very much. I was also good in mathematics.

After high school, there was no readily available opportunity for me to go directly to college. I took the examination for appointment to Annapolis, which was given when I finished high school. I didn't finish on top, so that opportunity slipped by.

In order to accumulate sufficient funds to go to college after I finished high school, I stayed out and worked. I worked full-time for Safeway Stores, which was quite an ordeal in those days, in the 1930s. You got $12 a week and worked six days a week. Saturday night was inventory night in the store, so on Saturday nights you'd finish up around ten or eleven. I think it was about a fifty-hour to sixty-hour work week at least, which wasn't bad. I never minded work. Then I left Safeway and got a job driving a dry cleaning delivery truck. I got $14 a week there, for six days a week.

Scott: Picking up dry cleaning?

Crandall: That's right. Picking up and delivering and so on. I got to know every street in San Diego by name and location.

Then I worked for a little local theater in my neighborhood, in the North Park area of San Diego. I was raised in that one area. We moved frequently. My grandmother felt that if you stayed in a house more than a year, something was wrong. I'd come home from school and find that we had moved. They weren't trying to abandon me, and always left instructions as to where they had gone. In any event, we moved a lot, but mostly always stayed in the same general area.

I worked for the theater, a small movie house, which cost only a 35-cent entry charge in those days. This work was much closer to home, had better hours and I made $14 a week. But I worked seven days a week there, doing every-thing from cleaning up in the morning to clos-ing the show at night. The only thing I didn't do was run the projector. That, of course, was beyond a kid of my age.

Scott: How old were you?

Crandall: About eighteen. I had finished high school, and for a while, that was it for education.

Chapter 2

San Diego State

Of the various science and math opportunities, civil engineering struck me as a good one, though I didn't really know much about that profession.

Crandall: After a year and a half of working, however, I decided that if I was ever going to college, I'd better make a break. I had saved up about $200, which I thought would help me. But just before I signed up to go to college, my grandmother had to have an operation, and my $200 joined the family funds to help pay for that. In February of 1937 I made the decision to start college anyway, and I am certainly glad I did then.

Scott::   Even though your grandmother had the operation, and that took your savings, you still managed to go ahead with school?

Crandall: I went on a shoestring, and went to San Diego State College, now San Diego State University, which was nearby. The fees were minimal. I don't remember exactly, but it probably cost $15 to enroll, something like that. I also worked Saturdays, which got me through all right, and I lived at home. We weren't starving—I do not mean to imply that. But there wasn't a great deal of money, either.

Choosing to Major in Civil Engineering

Crandall: I should say something about how I decided on my career. When I missed getting into Annapolis, I took stock of what I thought were my attributes and interests. It kept coming out that science and mathematics were the subjects that I felt I would like to be in and was good at. Of the various science and math opportunities, civil engineering struck me as a good one, though I didn't really know much about that profession.

Scott::   You mean, having some employment opportunity?

Crandall: I didn't think about a job so much, because at that stage of life I wasn't astute enough to check that out. But I did feel that I would like to be involved in building things. I had no basis for selecting civil engineering, other than what I thought my interests and aptitudes were.

At that time San Diego State only had a two-year program in the lower division, the freshman and sophomore years. I got started a semester early in the spring, which was very fortunate because I managed to build up a few units. I got good grades in things like freshman English and the history classes, a few things like that which were available, because I couldn't start any of the engineering yet. At that time you had to wait for the Fall semester to start the engineering. So I got about sixteen units of supplemental material under my belt,

which helped me later, because then I could take a lesser workload of electives. It also gave me a chance to get academically oriented and back into studying.

I started with engineering classes in September of 1937. I enrolled in engineering, which was a general course at that time, but you took mathematics and calculus, and surveying was also a requirement for civil engineering then. I took the surveying class, was proficient in it, and later got an assistantship helping the surveying instructor, looking after the equipment and assisting with the students.

San Diego State had some excellent faculty, who were very interested in their students' welfare, especially the math teacher, John Gleason, who also taught surveying. I had super courses in chemistry, geology, and physics. Those were good preparatory courses for me. In the sophomore year, you would begin to get into some engineering, mostly mechanical engineering, because the one faculty member who taught engineering subjects was a mechanical engineer. So we got some basic subjects out of the way at that point.

Summer Work

Crandall: In the summers I worked at the Safeway Stores again. They were very good to me. They didn't pay well, but at least I could get a job there. You were never exactly sure where you were going to be assigned. For example, I lived in North Park, the northeast portion of San Diego city, and they sent me to a store in Coronado. So I had to take the streetcar from home down to the ferry slip, transfer to the ferry, cross to Coronado Island, then take the streetcar again for another two or three miles into downtown Coronado to work. This is about an hour and a half trip each way. Safeway worked us long hours, so I was pretty well occupied just getting to and from work and doing the work. Later I was lucky and they transferred me to a store closer to home. I got pretty good in the produce department, working as a stock boy and that kind of thing. I enjoyed that work, and meeting people.

After finishing the sophomore year, I planned to transfer to the University of California at Berkeley. At that time, the only schools in California that gave a degree in civil engineering were Berkeley, Stanford, Caltech, and the University of Southern California (USC). Not even UCLA had upper division engineering classes as of then. So, having rather nominal financial resources, I chose Berkeley.

Surveying Class in the Sierras

Crandall: In order to graduate from Berkeley in civil engineering, you had to have two summer classes in surveying. The Cal schedule was different from the other schools—they started in August and finished in May. Since I didn't get out of San Diego State until June of 1939, it was too late to enroll in the Berkeley summer surveying class that year.

Fortunately, Fresno State College had a surveying class, called the Sierra summer school. We had about a six-week class at Huntington Lake in the Sierra Nevada up above Fresno. Most of the students there were from Fresno State, but there were two from San Diego State, myself and my roommate, Bill Brewer, who later went on to Cal with me. At Hun-

tington Lake we made a lifelong friend, Irvan Mendenhall, who is the Mendenhall in the architectural-engineering firm Daniel, Mann, Johnson, and Mendenhall. Irvan was also taking that surveying class.

When finishing my sophomore year, I was selected to receive an award from the San Diego chapter of the American Society of Civil Engineers (ASCE). It was the first student award they gave. I still have the picture that appeared in the newspaper at the time. It was for $25, which made the difference between me going or not going to the summer school class. I think one of the people who were involved in setting up that award was Paul Beerman, president of that chapter at the time. Without the award, I would not have had the cash to pay the fee for this summer school.

We had some exciting times driving to and from Huntington Lake. Bill Brewer, whom I've mentioned, had a Model A Ford, and we hooked up my father's little open trailer to carry our stuff. We drove from San Diego to Huntington Lake and had a couple of near misses and collisions.

Once, coming back down, Bill was driving, and he was unaware that we were on this steep grade, Tollhouse Grade, I think it was called. And this poor little Model A didn't have much in the way of brakes. We got started coming down that hill, and I thought it was curtains for us. Bill put the foot brake on, and I pulled on the emergency brake, and we got it shifted from high to second, and eventually down to low, and we finally pulled over to the side of this steep corkscrew road. We both changed our shorts and continued to drive home.

Chapter 3

University of California at Berkeley

It was the New Deal student job assignment that put me into the Berkeley soil mechanics lab.

Crandall: Bill Brewer and I, again using his Model A, drove up to Berkeley. We rented an apartment with two other San Diego fellows that we just happened to run into there. It was on the south side of the campus, and I remember that it was $25 a month for the four of us, in a two-room apartment. It had a kitchen, which we never really used because none of us cooked or cared about cooking. We didn't find out until after we had located a place to live that the engineering school was on the north side of the campus, and we were living four blocks south of the campus. So we had a nice little hike back and forth.

I was very fortunate and was granted a scholarship of $100 per year. It was a scholarship that a Holmes family had created in memory of their deceased son. The UC fees at that time included the registration fee of $27.50, and a laboratory fee for engineering and scientific courses of $17.50. So at that time, it cost me $45 a semester to go to Berkeley.

On top of that you had to buy your books and things of that sort. Most of the time I scrounged books from my roommates, or went to the library, but there were a few key books that I was able to buy used. That kept me pretty broke. I remember having only a nickel in my pocket for two or three weeks at a time.

We didn't spend very much. I went to the theater with the boys one time. I managed to get to two of the football games. One was when Cal played Michigan, and Tom Harmon was the big rage on the football field. That was the game when he was running away for a touchdown and some drunken person came out of the stands and tried to tackle him. Tom Harmon let him have a straight-arm and knocked this guy for a few loops.

I found a job washing dishes in a small restaurant run by a Greek fellow. I did dishes for my meals, many of which I didn't eat, because this was a real greasy spoon restaurant.

Scott: You didn't like the food all that much?

Crandall: The food wasn't that good. The most important thing on the menu was a rib steak for thirty-five cents. I worked my little butt off doing dishes. It was really a hectic atmosphere. The Greek owner would scream and swear at the help, not at me so much, but at some of the others. One boy there was a Jewish fellow, and the Greek guy would always malign him something awful.

Classes were interesting, and I did well. In the fall of 1939, I applied for a job with the NYA, the National Youth Administration, which at that time was the New Deal government agency that helped poor boys go through school. I think the pay was 40 cents an hour, and I was allowed ten hours a week maximum.

I guess it was a stroke of luck, but through no effort of my own I was assigned by the NYA to the soil mechanics laboratory, as we called it in those days, which was just getting underway. I think Berkeley had started it the year before, maybe in late 1938 or in the spring semester of 1939.

Scott: So the fact that you were randomly assigned to work in the soil mechanics lab is what gave you your first experience with what would later be your career?

Crandall: Yes. It was the New Deal student job assignment that put me into the Berkeley soil mechanics lab.

Professor Harmer Davis

Crandall: Harmer Davis was the professor of the graduate soil mechanics course. Harmer had been an outstanding student at Cal, and was then a very young professor. While he didn't like it, everybody but me called him "Stinky" Davis, after a cartoon character at that time. In order to look older, Harmer smoked a pipe and appeared very gruff, formal, and formidable, but he was really a nice guy. I got assigned to him. Harmer later specialized in transportation engineering and was chair of the civil engineering department.

I swept out the soil lab, which was just getting started. The soil mechanics lab had a corner assigned to it in the engineering materials laboratory building, in which there was some old cabinetry. I painted everything gray and did things of that sort. I helped out occasionally with some of the students taking the course.

We also had an engineer there working in the soil mechanics laboratory, not for the university but for the Bureau of Reclamation. A very fine man named Thomas Leps. Tom Leps was very, very friendly to me and helpful, and contributed much to my interest in soils.

I also made good friends with the other staff at the engineering laboratory. They had a machine shop there, and a bunch of really fine guys who were always playing practical jokes and things. A fellow named Joe Banville, who was called "The Scoutmaster," was the head of all the staff in the engineering laboratory. Under him was a very fine man named Eldon Whinier. Whit, as they called him, kind of took an interest in me, to the point where, when I was graduating and the senior ball was formal, Whinier loaned me the tuxedo he had worn when he was married, because I didn't feel that I could afford a tux for that. Incidentally, my fiancée had come up for the graduation, so I was going to the ball. These men were the people who built the experiments for the graduate students and took care of the big testing machine, the largest one in the United States I think.

Scott: What kind of testing was it used for?

Crandall: Materials testing for steel and concrete. It was about three-stories high.[1] Raymond E. Davis was really the head faculty person in the laboratory, and Davis Hall on the campus is named after that Davis, not Harmer.

In any event, back to the soil mechanics lab. I worked for the NYA ten hours a week, I think it was. We were limited to that because they wanted you to get your studies done. During the first year I swept out and did mundane things, but in the second year, Harmer Davis was designing an apparatus for compacting soils, and he put me on the drafting, which I frankly was lousy at. Not very productive. It seemed to take forever to get anything done, mostly because I'd have to pick it up and work on it for only two or three hours at a time, then put it away, then come back the next day and start over. But I developed an interest in soil engineering, or soil mechanics, which was the term they used then.

1. The testing machine is capable of four million pounds (18 meganewtons) compression and three million pounds (13 meganewtons) tension, and is still in use. A few years prior to the arrival of Crandall at Berkeley, the University acquired the apparatus to test large concrete cylinders, eighteen inches in diameter and three feet tall, with aggregate the size of baseballs—samples of material being used in the construction of Hoover Dam. After decades of service on the Berkeley campus, the machine was moved to the University's nearby Richmond Field Station and became part of the Earthquake Engineering Research Center there. It has been used in seismic testing to provide realistic simulation of large gravity loads on full-scale columns while lateral forces are simultaneously exerted by other devices.

Graduate Soil Mechanics Course

Crandall: When I finished my junior year and got to be a senior, I asked if I could take the graduate soil mechanics class. Harmer arranged it so that I could take the course in my senior year, even though it was a graduate course. I did well at it and got an A.

In those days, we had to do a thesis to graduate with our bachelor's degree. They don't anymore, I think. Two other fellows and I did our thesis on compacting soil. It was nothing earth shaking—an unintentional pun—but trod some new ground in the field of compacting soils in the laboratory. We put a lot of hours into the project.

Harmer gave me an A in that class. It was tough going, because about ten or twelve students were military people who had finished West Point and were taking engineering. They were going into the Corp of Engineers for the Army, and had come to Cal for a graduate degree. These guys were being paid to go to school. Also, most of them had a wife at home who cooked their meals and everything. So they were really hitting the books hard, at least it seemed to me. It was a tough class, and the grading on the curve was severe on most everybody who was only a regular student.

Scott: So your A was a pretty good accomplishment.

Crandall: Yes. Those were the days when an A was an A. While at Berkeley, the first thing I did was join the ASCE student chapter, even though it cost 50 cents. I managed to find that kind of money to join. Also, at the completion of my junior year I was invited to join Tau Beta Pi, the engineering honorary fraternity, and Chi Epsilon, the civil engineering honorary fraternity. Those memberships together cost $25. I went to the administration office and laid my financial position before one of the executives, and before I knew it, they had come up with a $25 loan for me to join these fraternities. They thought it would be a good thing for me to have on my record.

Seeking Employment

Crandall: Near the end of my senior year, Harmer Davis arranged for several of us who had taken the soil mechanics class to meet with a consulting engineer from southern California named William Moore, of Dames and Moore. Bill Moore came to Berkeley, and about three of us and Harmer met with him for lunch at the Faculty Club. It was the first time I had been to the Faculty Club, I might say.

Bill said he was looking for someone possibly to join their firm in Los Angeles. If we were interested he asked us to send a note to him outlining a little bit about ourselves and what we wanted to do. This was probably early May, 1941, toward the end of the senior year. I wrote him a letter. I still have the letter I wrote, from the Dames and Moore file. But time crept on, and I hadn't heard from Dames and Moore. I thought, "Well, that's not going to be a possibility."

So Bill Brewer and I and some others hopped on the train for Sacramento to talk to the State Division of Highways, as it was called then. Now it is called Caltrans. They were looking for engineers. The job market was starting to open up. Prior to this, engineers were hardly able to find any work at all. There was the war in Europe, and the United States was beginning to see that we had to do something, especially like supplying our allies through Lend-Lease, and other matters that would involve plant and facilities.

The Division of Highways had just made a change in their opening classification. Originally, you would start upon graduation with an engineering degree as a senior engineering aide, which paid $140 a month. Things had improved in the construction industry to the point where they weren't getting any applicants for that, and they upgraded the beginning position to junior highway engineer at $170 a month. Well, that made it a lot more interesting.

So I went to Sacramento and they offered me a job in the location I had asked for, my hometown of San Diego. I felt it would give me a chance to pay off some of my debts at the university while living at home. Besides, my fiancée, Eileen Exnicios, lived in San Diego. So I accepted the assignment and took the train back to Berkeley.

The next day I got a phone call. I was living in the two-story apartment building on Haste Street with about sixteen units. The phone was a common phone on the first floor. Our room had a buzzer. The landlady would answer the phone, and if it was for you, she'd give a certain buzz. So I got buzzed and went downstairs, and it was Bill Moore calling. He wanted to offer me a job.

Well, I was greatly anxious for that job. I was still in school and hadn't graduated yet, but I had already signed up for a job with the Division of Highways. I had told them "Yes," although I hadn't started work at all. Bill offered me the job, and I said, "Gee, I'd love to have it, but I've made a commitment to the Division of Highways and I have to stay with them." He said that he was sorry. So that ended that, at least for the time being.

California Division of Highways

Crandall: May 28, 1941, was graduation day. We finished school and headed back to San Diego. My folks had come up for the graduation. My grandmother, who had raised me, had passed away the year before, so she was unable to see the first member of the Crandall family finish college. I've always regretted that, but my grandfather was able to be there.

We came back to San Diego, and I started with the Division of Highways. That was the first of June of 1941. Eileen and I got married on September 20, 1941. Shortly after that, I became disenchanted with the California Division of Highways. They were all nice people, very, very friendly and kind, but I guess I wasn't cut out for civil service. I had interesting assignments. I participated in the design of one of the first cloverleaf freeway interchanges. That shows I'm getting old, because that interchange was torn down about twenty years ago. At the time, however, it was almost revolutionary for interchanges. They also put me in charge of the annual traffic count, where I worked under Ralph Luckenbach, who was a great mentor.

I did things that were very interesting and enjoyable, but for one thing we only worked 37.5 hours per week. You couldn't work overtime even if you didn't get paid for it, which we didn't. I wasn't used to just turning off the clock like that.

The other thing was that if you were designing anything, they had a manual. If you were designing a culvert, you just looked in the manual on the right page, and picked out whatever it was you were going to design. I figured I hadn't spent four years in school to copy something out of a book.

Moving to Dames and Moore

Crandall: So I decided to check with Dames and Moore again. Eileen and I drove up to Los Angeles from San Diego in our little 1935 Chevy. The car had trouble on the way, and I think we had to get a new clutch. At this time, there was the Los Angeles office of Dames and Moore, and Bill Moore was just beginning to

start the San Francisco office. It was late 1941. It was Admission Day, and the state people got a holiday, but other people had to work. I saw Trent Dames and Bill Moore there on that day, and they were nice, but they didn't feel they had any opportunities at the moment. So we went back home.

Then not more than a few weeks later Bill Moore called me one evening and said that they'd like to hire me. I said, "What are we talking about in pay?" He said, "How about $170 a month?" I said, "That's what I'm making here. It's going to cost me more to move up and live there." He hemmed and hawed awhile and said, "Well, we'll make it $175." That seemed like the world to me. Actually, I wanted the job. I'd probably have gone for less than $170. So I accepted.

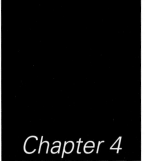

Chapter 4

Working for Dames and Moore

We started out trying to sell the discipline of soil mechanics to clients, and later had to sell the idea that they should hire us rather than all the other firms that started up.

Crandall: On December 17, 1941, just ten days after Pearl Harbor, I started my employment with Dames and Moore. At that point, nobody knew what the dickens was going to happen. Eileen and I moved up to Los Angeles and found a little apartment about four blocks away from the office, which was at Fifth Street and Figueroa Street, on the fifth floor of the Architects Building, which has since been torn down.

Bill Moore was spending most of his time in San Francisco. I don't remember if he had moved there yet or not, but they had plans for opening an office in San Francisco, and Bill was handling that part. I was in Los Angeles with Dames and Moore in soil mechanics and foundation engineering, as it was called in those days.

When I started, in December of 1941, they had a lead engineer working there, and I was under him. They had a couple of people in the laboratory, and about three or four out in the field taking samples and checking compaction. I think there was one secretary, and Trent Dames. At most, there were about eight or ten people in the office at the time.

Then in about March 1942, the lead engineer decided he wanted to do something else. The war was on. He moved into the shipbuilding business. I was promoted to the lead engineer in charge of the laboratory and the engineering functions. I was not chief engineer—Trent Dames was that at the time. I wasn't registered yet, of course, so I guess you'd say I was in charge of the office engineering function. Gradually, I got more and more responsibility. In 1944, I became registered as a civil engineer, which was about as early as I could get registered, because you had to have a certain amount of experience to qualify to take the test. I passed the test and pretty much ran the office.

Joining the Partnership

Crandall: In 1947, Trent and Bill offered me a partnership in the firm. They each were 50-50 partners, and each gave up some of it. I had 14 percent of the total, and later Bill Brewer, who had come to work for the company in San Francisco and was working with Bill Moore, became a partner, also at 14 percent.

I was the resident partner, they called it, in charge of the Los Angeles office from 1947 on. The Los Angeles office did more than just Los Angeles and southern California work. It was the lead office, so we also did the out-of-state work through that office.

Scott: Did Dames and Moore already have offices around the country?

Crandall: No, there were only two offices at that point, and Bill Moore hadn't really built up to a large office in San Francisco yet. But we did jobs all over, in other states such as Hawaii and even in other countries, for example in Holland and India.

Wartime Years

Scott: Before you go on, could I just ask you to talk about the wartime years. What was the impact of World War II on Dames and Moore and on the work you did—say during the period from the end of 1941 to 1945, '46?

Crandall: Practically all the work done was in connection with the military and defense. In fact, that was true of almost all construction—there was very little that was not in furtherance of the war effort. For example, we worked on the airplane factories, which were a big part of our work, and runways and military encampments. I remember Camp Cook, for example, up near Santa Barbara. War-related work in the San Diego area was important. We did a lot of paving evaluation in those days. They were just beginning to come in with the heavier planes, such as the bombers, and the original airfield pavings were not standing up. So what we were doing then were the plate load tests, bearing tests. We'd get a big truck with some load on it, and put a jack between the axle and the ground on a plate of a certain size, and run a test. The California bearing ratio test was devised just about then, too. So the evaluation of existing runways and of new runways was becoming prominent. Until then, they'd been designed by the seat of the pants. We did the soil consulting engineering for the Douglas Aircraft plant in Long Beach, which was built at that time. Lockheed and North American Aviation were going strong. All of those facilities were requiring soil engineering—fortunately for us, since there wasn't anything else to do.

One of the things that disturbed me was my feeling that I needed to contribute something more to the war effort. I applied for and was promised a commission in the Navy Seabees, to go overseas. They were building airfields, working with heavy equipment, and so on. I thought I would be a good addition there, and would get some good experience. A faculty member of the University of Michigan I believe it was, named Bill Housel, who was a commander or captain or some type of officer in the Seabees, was putting together a group.

I contacted him to see if he could use a soil mechanic. He pulled a few strings. I applied, and while I had hoped for more, they offered me an ensign commission in the Navy, with the understanding that I would be assigned to this kind of work, with Housel's group. It was practically consummated. I had done everything but sign on the line.

Prevented From Enlisting in the Navy

Crandall: I told Trent Dames what I was doing, that I thought I could be of more value to the war effort in the service. He contacted the draft board and told them what kind of work we were doing at Dames and Moore and how important that was, and the draft board issued a telegram. I got a copy of the telegram they sent to the Navy saying they wouldn't let me

go. They felt I was more valuable in civilian work. So I didn't go.

At the time I was very unhappy about it. It turned out, however, that this was one of the greatest things that ever happened to me, the fact that I didn't go into the Navy. It was getting near the end of the war, and the guys who had been in the longest, justifiably were being released the earliest. They had the points, you know. I would have gone in at the end, and would probably have sat at a desk somewhere for several years after World War II.

Scott: Did Trent Dames take these steps on his own?

Crandall: Yes, he did. He didn't tell me about it until the word came through. I almost quit, I was so upset about it. You couldn't leave a job in those days, and you couldn't get any raises or anything. The job market was completely frozen.

Scott: I guess Dames felt morally justified, in that he believed he had more important business for you to do?

Crandall: Yes, and he didn't want to lose me. I guess I was pretty good. You could hardly get engineers at that time. If they graduated through the V-12 programs,[2] or whatever else

---

2.   The V-12 Navy College Training Program, begun in 1943, was designed to provide college-educated officers for the war effort, contending with the problem that the draft age for males was 18 and prevented them from attending or graduating from college. V-12 students were technically already in the service and underwent some military training while studying year-round. Upon graduation, V-12 students were sent to Navy or Marine training programs and became officers.

they were in, they went right into the service. I was pretty much responsible for running the whole darn Dames and Moore shooting match in Los Angeles at that time. Even then, Trent had big ideas about starting an international organization, and opening offices in the rest of the world.

Scott: And he didn't want his local organization falling apart while his attention was focused elsewhere?

Crandall: He wanted to be free to plan these other things. It was the way he operated. I was very angry at first, but I saw the war was winding down. The atomic bombs were dropped on Japan about that time, ending the war. So it turned out to be a good thing for me, although I never did get to be in the Navy. As I mentioned earlier, in high school I had aspirations of going to Annapolis, so the Navy was still a matter of interest to me.

Postwar Work

Crandall: Now I must relate a story that happened after the war, when civilian work was starting up again. Dames and Moore sent me on a business development trip around quite a bit of the United States. I remember going to New Orleans, Houston, St. Louis, Chicago, places like that, to sound out the attitude of people toward soil engineering. Later this was fed into Trent Dames's mental computer as to whether or not he would try to open an office in some of these areas.

In those days soil mechanics was brand new. Dames and Moore was something like the second or the third firm in southern California

to do soil work. Most engineers and architects thought it was a lot of baloney. They had designed foundations for years by going out and pushing their heel in the ground and saying, "That's good for 4,000 pounds" [4,000 pounds per square foot] or getting information from a building next door and applying that.

Scott: In those days, then, only the part of the structure from the ground up was considered important?

Crandall: Yes, from the ground up. The attitude was: Who cares about foundations? You just poured concrete into the ground and it usually behaved okay.

I'll never forget one experience I had in Seattle. I would go through the phone book and find architects' names and addresses, and engineers, and structural people, and then beat on their doors and try to tell them how great soil engineering was. Shouldn't they be interested in a soil engineering, soil mechanics firm?

Well, this old architect listened to my story, and finally said, "Listen, son. I don't know anything about your business at all. But I know this. Ninety percent of the buildings are held up by friction and the grace of God." He's pretty close to right, I think. That always stuck with me.

But one older structural engineer in Chicago, who listened to me patiently and was trying to sweep me under the rug, finally told me what he did. He had the Raymond Concrete Pile Company, which had a drilling business, go out to the site and take soil samples in their little sampler. They put the results, what he called "rat turds"—they were pretty good-sized rat turds—in a glass bottle. The Raymond Company would send the bottles to this guy, to see what he wanted to design his foundations for. He had a roomful of these cardboard boxes with these little bottles of soil samples.

He said, "Yes, I just take the sample out and squeeze the soil and I decide how good it is." Then he thought for a minute and said, "You know, there's one thing, though, I guess I'm getting weaker in my old age, because I'm giving higher values now for the same soil. Something that ten years ago I would give 10,000 pounds to, now I'm giving 20,000 pounds." He then said, "Maybe we do need something a little more scientific." But he didn't hire me for anything.

Scott: How often did you do these tours?

Crandall: Just periodically. It must have been about 1950. The war was over, of course, and I was a partner at that time. I left Dames and Moore in 1954, in mid-year 1954. So it had to be probably between 1951 or 1952, somewhere in there.

Selling the Discipline of Soil Mechanics

Crandall: At that time the problem was to sell soil mechanics as being an important element of design and building. Nowadays, soil mechanics is accepted. Hardly anything is built, at least in southern California, without soil mechanics. We started out trying to sell the discipline of soil mechanics to clients, and later had to sell the idea that they should hire us rather than all the other firms that started up.

Scott: There is now an awareness that things can really go wrong if designers do not understand the performance and the weight-bearing capacities of the soil?

Crandall: You had to get across the idea that we could provide some useful information. Because most people, when you come along and try to sell a new concept, they feel they have gotten along fine with what they had before. In effect, you're telling them, "You're not doing things as well as you should have been doing them." It took a good man to stand up and say, "OK, let's see what you have. Maybe you can teach me something."

Trent Dames and Bill Moore

Crandall: Dames was not the salesman type. He was the administrator. A good technical man. He loved management. A lone wolf. He alienated more clients than he got, frankly. I think he realized that, and he got out of the way of client relations. I guess I was better handling or dealing with people than he was, so client contacts became one of my assignments.

Scott: Operating out of Los Angeles.

Crandall: Yes.

Scott: Bill Moore also did a lot more client contact work, I guess, but he was operating more out of the San Francisco office about that time?

Crandall: Bill also went over to Saudi Arabia for Standard Oil of California and Aramco, the Arabian American Oil Company, and worked at getting them underway on building refineries there. It's mentioned in Bill's

EERI oral history.[3] He shipped the samples back to Los Angeles and we did the testing. We worked out a code system for sending him data from Los Angeles. When he got testing equipment over there, Bill ran the tests and by code would send the results of the test. We'd draw up the logs and other things from the data that he sent. It was expensive and difficult to get messages back and forth. We tried to keep it short and sweet. Aramco, I guess, was able to get the code across by radio telegraph, or something like that. Also, Bill was busier than a bird dog in San Francisco doing his own business development. Up there the architects and engineers were a lot more provincial and less progressive than they were in Los Angeles.

In 1947 I became the first partner of Dames and Moore. We were a partnership then; it wasn't a corporation. I had responsibility for what was called the Los Angeles regional office. At that time there was also an office in San Francisco, and Bill Moore was heading that. Trent Dames was in what we would now refer to as the corporate office, but back then it was called the general office.

The general office took care of the total business picture, and was working toward establishing other offices in other areas. Later this was done in New York, Seattle, Portland, and areas like that. It was the main function of the general office. The Los Angeles regional office was responsible for all of southern California,

plus much of the foreign work that we did. That was my responsibility in Los Angeles.

Postwar Surge in Civilian Construction

Crandall: Following the war, in about 1947, the basic work was commercial, filling a demand for the buildings that could not be built during the war. There was a great surge of private work, as compared with the previous defense and war-related work. We were very fortunate in being able to move almost without any hitches from government defense work to private work. In other words, all our eggs were not in one basket anymore.

Scott: You didn't have to retool the office?

Crandall: Right. The main effort was on schools and municipal public buildings, whose construction had been curtailed during the war. Then, of course, there was the general private sector, such as buildings for the telephone company and gas company.

Refineries

Crandall: Refineries were beginning a big building program about that time. We were doing much work with groups like Union Oil—and what was then Richfield and later called ARCO—throughout not only southern California, but throughout the southern United States. We did work in Texas, for example, and also in Kansas City, Chicago, and other areas of the United States, for refineries designed and built by C.F. Braun, in particular. Headquartered in Alhambra, California in the Los Angeles area, C.F. Braun & Co. was one

---

3.   *Connections: The EERI Oral History Series—William W. Moore:* Stanley Scott, interviewer. Earthquake Engineering Research Institute, Oakland, California, 1998, p. 35.

of the outstanding petrochemical design-and-build firms at that time.

We had a fine relationship with C.F. Braun, and did the soil engineering work for all of their projects throughout the United States. We must have done fifteen or twenty major refineries with C. F. Braun. Unfortunately, they are no longer the lead. Some years back they were acquired, and the character of their business has changed. Nobody's building refineries nowadays, but at that time, they were a major part of our work, involving large projects around the country.

Coastal Facilities

Crandall: There were also some interesting offshore projects. One example is the Hyperion sewer line, an interesting project. We did the soil study for the five-mile outfall sewer that extends off the Los Angeles coast.

Scott: That must have been a major project.

Crandall: And a much-needed development. The effluent was taken five miles out to sea, and then everybody thought it was fine. Nowadays, however, they have found that sludge accumulation is causing problems, and they're doing things a little differently.

We did several offshore projects of that type. A number of piers were built, such as the Venice pier. In San Diego, there was an offshore outfall, and two or three other piers. So marine work was going strong, including harbor department work in Los Angeles and Long Beach. There was a backlog of development to be done because of the war.

San Diego began to boom, and we were the prime soil firm in connection with major developments in San Diego for the Navy. Many of their shore facilities had been limited during the war, and they began to expand. Those are just a few projects that come to mind at the moment as being rather noteworthy in that period. They gave us the chance to expand our techniques and knowledge into other areas.

Development of the Drilled Friction Pile

Crandall: The drilled pile was one development in those early days that was a very important foundation technique. A hole was drilled into the ground, and then filled with concrete.

Scott: That was done, instead of driving the pile down into the soil with a pile-driver?

Crandall: Yes, instead of driving a pile into the soil, which was one of the standard procedures used for many, many years. In southern California, a drilled pile was often very economical, and if the conditions were right, much more economical than a driven pile. Of course, where there are sandy soils with shallow water conditions, the hole won't stay open, making a drilled pile hard to install. But where there are dry conditions, and the surface materials are not suitable for conventional spread footings, the drilled pile was the answer to a maiden's prayer.

This development was pretty much pioneered in southern California, using what was then called a cesspool rig. It was used actually for that purpose, drilling cesspools. They drilled a hole in the ground that they lined with redwood. For the drilled pile, that same bucket-type drilling equipment, as it was called, was used to drill a hole in the ground, and concrete was put into the hole, and a friction pile developed in that fashion.

It had to be proven to people that you could develop friction on the side of a hole that was merely filled in with concrete. They thought that to develop friction, the pile had to be beaten into the ground. So several tests were made in the early days, some of which were prior to my even coming to Los Angeles. They put a drilled pile in the ground, and then drilled another hole alongside it, say five or six feet away, both to the same tip elevation. Then they tunneled under the drilled pile to remove the soil from beneath its tip. The pile was then loaded to prove that the tip was not carrying the load—instead it was the friction on the sides that carried the load.

Once that principle was established, people began to believe it. Most engineers had felt that the load was going all the way down to the tip, and that you can only load that pile up to whatever the eighteen-inch diameter tip area would support. But that was not the case, and the drilled pile would take considerably greater loads than that. The development of the drilled, cast-in-place concrete pile was a pioneering effort in southern California. We got that type of foundation permitted in the building code, and in this area a very large number of buildings have been put in that are supported on this kind of piling.

Parting with Dames and Moore

Crandall: I had a very fine relationship with both Bill Moore and Trent Dames. Things went very well, although there were some business differences. Trent Dames was internationally minded, multi-office oriented. In that regard, I had some reservations, purely from a business standpoint. We had several discussions about what was going to be done and how.

Scott: You had reservations about the opening of other offices, or soliciting lots of work away from home base?

Crandall: I was concerned about possible effects on the quality of the work, if offices were opened without suitably trained personnel. Trent had different ideas on that.

Scott: In other words, he thought the quality problem could be handled, or he was eager for the business, or maybe both?

Crandall: Expansion was his middle name. Bill Moore, I think, was less oriented that way. But at that time, in the early 1950s as I recall, there was an executive committee consisting of myself, Bill Moore, and Trent Dames. The matter was discussed in the committee, and it was obvious that I was not in tune with what the others wanted to do.

Scott: You felt it would be better to stick with two or three offices.

Crandall: Or at least increase in size more gradually. This was a partnership, you will recall, and each partner was vulnerable for whatever happened in any other office. One of my concerns was that, if we opened an office in another area, we could all be affected if that office developed some liability problems. Under a partnership, those problems could affect us all.

Here I was, a fairly significant partner in the legal entity of Dames and Moore. While not rich by any stretch of the imagination, anything I had would be exposed to whatever actions or claims the work of the people in other Dames and Moore offices might generate, even by unintentional things. So I was less enthusiastic about being, shall I say, in bed with other office managers over whom I had very little control, and about whose abilities I had little knowledge.

I recognized that my thinking ran contrary to what Dames and Moore were contemplating, and in fairness to both them and myself, I felt the time had come to separate from the firm.

I also sensed that there was something of a corporate bureaucracy developing, in which headquarters staff tells all the workers what to do. The line personnel, the professionals, can become secondary. These things didn't fit my idea of how to run a company. Not that there's anything wrong with that model, and Dames and Moore went on to become one of the largest firms of this type in the country, maybe even in the world. That's fine—but it wasn't for me. I didn't see that approach as being my cup of tea. In early 1954 I concluded that what I was doing was not good for either Dames and Moore or for me. It wasn't right to them, to have somebody who was not actively pursuing the policies the company was interested in.

The final decision was made in 1954. I had an offer from another firm, primarily a testing laboratory, which wanted to do soil engineering. It was located in my old home town of San Diego, where my wife and I had been raised, and where we had our family (my brother, sister, and parents). So we thought about moving there. I regret to say I had practically assured the other party of our intent, to the point where he was kind of counting on my coming down there. At this time I had advised Dames and Moore that I was planning to leave, and that I had been asked to come to work for the San Diego firm.

Meanwhile, some of the structural engineers in town who were good friends of mine, and with whom I had worked as a Dames and Moore partner, convinced me that I would be better off to stay in Los Angeles where I had all these contacts. They all felt that my services would be preferred to whoever else might come and take over at Dames and Moore. I shall not name them, but there were two in particular.

Scott: Two structural engineers, friends of yours in this area?

Crandall: Structural engineers, yes, who very strongly urged that I should stay in this area.

Scott: Were they your peers, chronologically?

Crandall: Somewhat older, but only by four or five years. Then two of our associates, employees at Dames and Moore, Leo Hirschfeldt and Fred Barnes, approached me, saying they were interested in working with me if I started my own firm. Neither of them was a registered civil engineer, but they had been with Dames and Moore almost as long as I had, and one even longer. They offered to join me in forming our own office in Los Angeles. We agreed that this might be a good thing to do. I had to tell the San Diego contact that I had changed my mind. He was very gracious about it, but he wasn't very damn happy. I've always felt I left him holding the bag.

Financing was obviously a problem in starting a new business. All of my assets were tied up in Dames and Moore, in ownership and in retained earnings. In Dames and Moore we operated on a drawing account, which, I think, at that time was $600 a month. That's what you lived on. Then theoretically at the end of the year after the books had closed, if there was a profit, it was distributed. Initially, 14 percent was my share, but later this was changed. We decided that we could probably last for six months if we could come up with $10,000.

Scott: You mean $10,000 cash total, for the three of you?

Crandall: Yes. We raised this nest egg from $5,000 paid in by me, and $2,500 from each of the other two. We'd see how it went. If we made it, we made it. If we didn't, it was down the tubes and we'd do something else. We were kind of lucky in our timing. By then I had advised Dames and Moore that I was resigning. I think I gave them a three-month period during which I would stay on, and work with and train my replacement, a very fine fellow named Al Smoots, who was going to take over the office. I did stay, and left about May 1, 1954.

Prior to my finishing, but after the decision had been made that I would go, these two other parties approached Trent Dames, told him they were planning to leave, and gave him a month. We felt the firm deserved a month notice. Fortunately for us, Dames decided he didn't want them around if they were going to leave, so he terminated them right then. That timing turned out to be the best thing that ever happened, because we started getting work the day we opened our doors. Without them I would have been the only one to do all of this work. So the other two came aboard right away. It worked out very well.

Scott: Was your parting with Dames and Moore amicable?

Crandall: It wasn't really amicable with Trent Dames. In fact we had a financial dispute over the payout of my share in the firm that dragged on. I can say to this day that my firm did not go and solicit a job from someone who had been a client of Dames and Moore. We did, of course, send out announcements about our firm, and we had it put in the magazine that LeRoy Crandall and Associates had been formed, and that sort of promotion. We ended up getting a big job in the mid-1950s for further work on the Hyperion sewer outfall, but only because the city engineer had strained relations with Dames and sought me out for our services.

Chapter 5

LeRoy Crandall and Associates

We started the office with a card table, a desk, a second-hand typewriter, one three-drawer metal file cabinet, and a drafting table.

Crandall: We opened our new firm's doors in 1954 and had work before we were even ready. We didn't have our testing machines and other equipment. People came in and called us and wanted us to do work. We never had any problems with obtaining work.

Scott: Your structural engineer advisors had called the shots pretty well in recommending that you stay in the Los Angeles area.

Crandall: Yes, they were right. Many people helped us very much. One very, very fine engineer, Jim Montgomery of J.M. Montgomery Engineers, called early-on. One of our first jobs, job number eight it was, was a reservoir in the Las Vegas area. Jim called and asked if we would do the work. I said, "Gee, Jim, we'd love to. I don't know if we can finance it, though." It was a pretty big job for us, about 30 borings and things like that, maybe it was a $15,000 project, which was a big fee for us. I said, "Financing might be a problem." He said, "Look, LeRoy, don't you worry about that. We'll pay you in advance. We want you to do the job." We had support from people like that, which really made it worthwhile.

Scott: He offered to pay in advance?

Crandall: We didn't need it if things worked out all right, but we had only $10,000 total, and we had to hire drilling equipment, and pay them to operate it.

Scott: You had a cash flow problem.

Crandall: Check. One thing that I was proud of, and I'd like to beat my drum about, was that the first thing I did when we opened the doors and took in some hard cash was to join the U.S. Chamber of Commerce, the California State Chamber, and the Los Angeles Chamber. We've been supportive of things like that right from the start. For whatever that's worth, I felt we were here to stay, and we were going to make a business and do our share of trying to support private enterprise.

Like I say, things went well. I can't recall any major problems. We grew from the three of us engineers and my wife, Eileen, the secretary. We started the office with a card table, a desk, a second-hand typewriter, one three-drawer metal file cabinet, and a drafting table. We rented a little office space on Beverly Boulevard in Los Angeles.

Scott: You started almost on a shoestring.

Crandall: Oh, we had sandals without shoestrings, I guess you could say. But the $10,000 was adequate, along with the kind of support we got from our clients, who paid quickly and well.

Hyperion Sewer System Expansion

Crandall: As I mentioned earlier, one of the big jobs our new firm had was the expansion of the Hyperion sewer outfall system for Los Angeles, a project on which Dames and Moore were consultants during its first phase. The city engineer, a man named Aldrich, preferred not to use them again, and although the project was a stretch for our small office, we took it on, to do the soil report for the new outfall sewer. I think the total fee was something like $75,000. That's not so big by today's standards, but back then it was a hell of a big job.

Scott: I remember the Hyperion outfall debate and project, which was news even up in the San Francisco Bay Area. It got into some of the literature that came across my desk in my early days at the Institute of Governmental Studies. As I recall, that project was a very important issue at the time.

Crandall: The outfall sewer went out six miles. We also had to drill some other borings to check the sewer installation and find out the soil conditions. Then there was about 15 miles of onshore sewer line at quite a depth. Much of it was put in as a tunnel, tunneling under the Los Angeles airport and that whole area in Baldwin Hills. It was quite a job and a feather in our cap, believe me. We did it well.

Organization of the Firm

Scott: How did you organize the firm?

Crandall: The other two partners, Leo Hirschfeldt and Fred Barnes, had a quarter interest, and I had half. We selected the name "LeRoy Crandall and Associates," which doesn't show much originality, I guess, but it seems to have worked. That was the best selling approach, I think, because I was better known than the others.

A few months after we had opened, Russ Weber, who also worked at Dames and Moore, came by. Russ had approached me earlier about starting a company, but at that time I had already committed with Fred and Leo. But Russ was now ready to join, and we took him in as an equal partner to Fred and Leo. They each had equal shares and mine was twice their individual shares. That comes out 40-20-20-20—that is, they each had 20 and I had 40. We took off on that basis.

I would say within about three or four months we had increased our staff from the four of us to about seven altogether. Our efforts were in the Los Angeles area, of course, and purely in soil engineering. That was the whole thing at that time. The field enlarged in later years, but initially it was just exploring and testing the soil for foundation design purposes.

Crandall and Associates grew in size. I always had felt that about sixteen total personnel would be what I considered ideal for a small consulting firm that believed in quality service. That was about the size I had in the Los Angeles office of Dames and Moore, about sixteen or eighteen.

In the new firm, however, we got to that level fairly quickly. I don't have the numbers in front of me, but I guess within two years we were up to that size of total personnel. I think my theory was good—I still believe that is a good size for a principal to operate and conduct a

business up to the point where you still know almost everything that's going on.

Scott: Being intimately knowledgeable about every job you are doing?

Crandall: Yes, that's it. And that was the basis for my leaving Dames and Moore, as I said. Also at that time we were beginning to be conscious of liability, of lawsuits against soil engineers. I was no genius at management, but I was smart enough to recognize that if you didn't have good controls, you could easily get yourself into a legal situation that should otherwise have been prevented, if you had known what was going on in time to take some action before problems developed.

Well, the theory was good, but we couldn't hold to it. That was not because we were out soliciting every job that came along. But we had developed a clientele of, I will say, the best architectural and engineering firms in southern California who relied on us almost automatically for their soil work. They grew as Los Angeles grew. And when a firm such as Daniel, Mann, Johnson, and Mendenhall gets bigger and has more work, and they want you to do their work again, you'd better be prepared to do it properly, or they're going to look for someone else and you won't have any work. That's exactly what happened, so we had to keep growing.

Scott: That's interesting. There is pressure on you because of your success and your clients' success. I gather you almost can't escape it.

Crandall: That was exactly right. We were very slow in soliciting new clients, because we had the cream of the crop and they developed all the work we needed for a small-size firm. But they would grow, and then, of course, as people left those firms and started their own firms, you would then have two or three organizations that still looked upon you as their consultant in this field.

Incorporating in 1957

Crandall: When we started, it was a partnership. Then after about three years we incorporated, and each became stockholders. There were some benefits to that. If I remember correctly, Leo Hirschfeldt searched around and found that you could operate under what was called "Subchapter S," which permitted you to divide up the profits as if you were a partnership, but gave you many of the benefits of incorporation.

Scott: Also, I gather, it freed you from some of the vulnerability of a partnership.

Crandall: That is right. At that time at least we thought—later it wasn't quite as important—but a corporation was less vulnerable then. In a corporation, the individuals were less vulnerable in the event of a lawsuit or some horrible catastrophe. Later litigation indicated that they could "pierce the corporate veil" as lawyers love to say. If you're a professional person and had signed drawings and stamped the drawings with your registration you could be held liable as an individual. So it didn't have all of that reduced vulnerability aspect for very long, but tax-wise it was a good move. Later, we got so big that we had to drop Subchapter S. I've forgotten now, but you could only have X number of partners or

stockholders or whatever. There was a lot of legal mumbo-jumbo about it.

So we had to expand. Expanding meant larger quarters. We rented or leased space for a while, and had it added onto. Then, I think in 1965, we decided to build our own building. I was not really strongly in favor of that, because I felt that we were better off to keep our money in our own field and let somebody else own the building, but it turned out to be a damned good investment. I think it was 1965. That would have been 11 years after we started.

Scott: In hindsight, that would have been a good time to build or buy, seeing what happened to the real estate market. It really took off, starting in about 1964 or 1965.

Crandall: Yes. It was dumb luck. Leo Hirschfeldt was the one who maneuvered us into that. Leo was more of a business manager than a civil engineer. He was a graduate civil engineer, but he never got his registration. He loved the business aspects.

We hired one of our architectural clients, a fellow who had been with a big firm and left, and he designed what I felt was a very, very fine building for us to operate out of.

Scott: Where was it located?

Crandall: At 711 North Alvarado Street in Los Angeles, near Echo Park Lake. The neighborhood was not very classy, but we felt it was going to improve. That was the one thing we were wrong on. It didn't improve much while we were there, although now it has, after we've sold the building. We built that building at the intersection of Alvarado and Kent Streets, so we called the corporation Alvarado-Kent

Corporation. We built a 10,000- or 12,000-square-foot facility that included storage for our equipment, a laboratory, and our engineering offices. It was one-story high, but we had designed it for a second story, so we could expand if we wanted to.

By that time I think we had six partners. Jimmy Kirkgard and Seymour Chiu were the additional partners we had added. We called them associates. Both of them came within roughly a year after we had opened the door on the new building. Seymour was from Hong Kong and had a master's degree from the University of Texas. Jimmy Kirkgard was a UCLA graduate, with a master's degree. Martin Duke sent him over to us.

Each of us invested in the building in an amount equivalent to our ownership. We felt Jimmy and Seymour had the abilities and the talents and the qualities that we wanted, so we offered them a share of the business, which they accepted. They bought their own stock. We didn't give any stock away. We permitted them to buy into the business. So when we built Alvarado-Kent, six of us had shares in the building—me, Fred Barnes, Leo Hirschfeldt, Russ Weber, James Kirkgard, and James McWee. Later on, Seymour Chiu, who had been with us for as long as anybody, was made the seventh associate, and after that Perry Maljian was selected as the eighth.

I felt very strongly that we should make the business available to our key personnel, if we expected to keep them. If you get good talent, unless they are "part of the action," they're going to leave after a relatively short period of time, after they've achieved everything they're going to get. If they're just working for a salary and a bonus, it isn't nearly as interesting as having a portion of a business that they can devote their time to.

Scott: You mention C. Martin Duke. Had he worked with you before?

Crandall: Yes, he was by then a professor at UCLA, and we were very close, and we collaborated on a couple of things. I'll mention him several times here in this oral history. I have forgotten whether we hired Martin and he got paid, or he was working on a research grant, but he did early shear wave velocity measurements with us.

I guess we moved in 1966. We finished the building in less than a year. We had a twenty-year mortgage, and paid it off in ten years, so then we owned the building free and clear. Crandall and Associates paid rent to Alvarado-Kent. That turned out to be a financial blessing because when we finally sold the building [in 1986], we sold it for a million dollars, and I think in 1965 we had paid a couple hundred thousand, something like that.

Scott: Why did you sell?

Crandall: It was related to the next phase in the Crandall firm, when we were acquired by Law Engineering.

Acquired by Law Engineering

Crandall: Law Engineering of Atlanta acquired Crandall and Associates in 1982. At that time, our firm had seventy or eighty employees. The name "Law" comes from the firm's founder, Thomas Law. We were to participate in the earnings or profits over three years, between 1982 and 1985, with a maximum value equal to the amount they paid for LeRoy Crandall and Associates. LeRoy Crandall and Associates became a subsidiary of Law Engineering Testing Company in 1982 but retained its former name until 1991, when the name changed to Law/Crandall. Law didn't buy the building, so Alvarado-Kent still owned it.

In 1985 the earn-out period ended. Law has a growth policy. They wanted to be big. So we outgrew the building, and Alvarado-Kent offered to add on the second story. But the parent company, Law Engineering, decided that rather than stay in the building while the second story was being added, they would move out and lease larger quarters. The decision to lease larger quarters left Alvarado-Kent with a building to dispose of, since we didn't feel like trying to lease it out. By that time, Leo Hirschfeldt and Seymour Chiu had passed away, Fred Barnes had retired, and Russ Weber was just about to retire. So rather than try to keep the building, we decided to sell it in 1986. LeRoy Crandall and Associates then found quarters in Glendale that were much larger than what we had before. The move to Glendale was made in 1986. Then, due to a need for even more space, in 1991 another move was made to a newly built two-story structure in the City of Commerce.

Geology and the Practice of Crandall and Associates

Crandall: This brings me to a point about the evolution of my firm. Crandall and Associates limited ourselves to soil engineering, and would use consultant geologists when we needed that type of input. For some time we did that, using two firms for the geologic work, James Slosson, and Glenn A. Brown and Associates.

Then it got to the point where many of our competitors had in-house engineering geologists. Also, some of our clients expressed a preference for a firm that didn't submit two reports, but would combine both the geology and the engineering in one report. We used to have a report written by our consultant, let's say, Glenn A. Brown, and we would append that geology report to our soil report. We, of course, used the information from it, but two separate reports were sent to the client.

So in order to meet our clients' desires, LeRoy Crandall and Associates merged with Glenn A. Brown and Associates. Glenn Brown and his staff, about ten or twelve people, became part of LeRoy Crandall and Associates. Our firm before that time was about forty or fifty people. Glenn Brown was brought in as an associate of the company. That was in the mid 1970s. Glenn had a very fine reputation, and we were very fond of him and his work.

We worked out fine with Glenn Brown. We acquired his firm, his equipment and apparatus. He acquired stock in LeRoy Crandall and Associates, and became another co-owner. The firm then became an integrated operation, and we identified ourselves as "Geotechnical Consultants," rather than just "Geotechnical Engineers." This included the geology that Brown was in charge of under that broader designation.

Chapter 6

Development of Soil Engineering

The model grading ordinance of the City of Los Angeles and the tie-back anchor are widely used foundation and soil engineering approaches that were pioneered in southern California.

Convincing Them We Had Something Useful

Crandall: I think most builders and design professionals thought we were like the guys with the water witching techniques. It was all mumbo-jumbo—who needed all that stuff? It was a question of convincing people that by taking samples and running tests and doing engineering analyses, you could develop good, useful information.

At the time, the Navy, Air Force, and Army seemed to be convinced that there was some merit in this sort of a thing. Then people began to find out that when things went wrong with foundations, it was considered the designer's fault. But now here was another layer that would step in and assume responsibility for the uncertainties in construction that are primarily caused by the underground conditions.

With responsibility went respect. So we began to get respect, at least as representing a buffer, separating the designer from some of the problems of construction. The designer then had the soil engineer to take the brunt of the attack if anything went wrong with the site or foundation. Basically, it was just a question of their changing views. Previously they had gotten along, for thousands of years, without running soil tests. So the attitude was "What have you got that's going to be helpful to me?"

That skepticism gradually changed, as building departments began to rely on the soil engineer's findings, and owners discovered that they could save considerable money by knowing the exact design problems on a site, rather than just arbitrarily applying the prescribed design values in the building code.

Scott: In other words, they could tailor-make what they did in terms of foundation and preparatory work?

Crandall: That's exactly right. You learn the conditions of the site, and since building codes are conservative documents, usually you can save money. In other words, in most cases the actual soil value determined by the consultant is better than what the building code requires if you don't have a site-specific study.

You might be able to design the foundation for, say 5,000 pounds per square foot bearing pressure instead of the 2,000 pounds that the code might otherwise say was the presumed value. A great deal of thanks is owed to the building department people of the City of Los Angeles

for realizing that early-on, and for writing the code in such a manner that deviations from it were permitted on the basis of a qualified soil engineering report.

Many of us in the soils field were instrumental, working with the Los Angeles building department, in getting the information in there. So it became possible to deviate from the building code on the basis of an acceptable soil study.

Leadership by the City of Los Angeles

Scott: In this respect, the City of Los Angeles has tended to be a little ahead of the game?

Crandall: I think there is definitely that factor—not only in the quality of their plan-checking department and personnel, but also in their acceptance of new techniques and allowing for those not specified in the building code. Then, of course, they've also tightened up many things, for example the grading ordinance, which the local soil people also helped develop.

Back in the 1950s and early 1960s, I think the City of Los Angeles certainly led the United States, and maybe the world, in requirements for evaluating soil properties before constructing hillside developments. Because of the previous lack of controls over the developers, in our hills there were many problems with stability, landslides, and erosion failures.

It's always struck me that in other parts of the world, the very poor people live on the hillsides. But here, in southern California, a hillside lot is a desirable place. So we have concentrated a great deal of expensive development on hillsides, and as a result, hillside development tends to be controlled in ways that minimize grading problems. I can now say that the hillside area of the City of Los Angeles has fewer problems than almost any comparable place I can imagine. The proportion of trouble is very, very minimal.

Scott: I think this is a very important point—that the City of Los Angeles often seems to be a little ahead of most of the rest. Do you have any thoughts as to why this was the case? Is it due principally to professional leadership in the appropriate departments? Do you have any general ideas as to why Los Angeles is often a bit ahead of the game?

Crandall: I suspect it is that we have a vocal citizenry, who have built expensive homes in the hillsides. When trouble developed when heavy rainfall occurred in the early 1950s, there was a great outcry. "What's wrong with our hillside development?" Good soil engineers already knew what should not be done, but there were relatively few controls on the developers.

The 1952 Los Angeles Grading Ordinance

Crandall: It was just a question of slope requirements not being appropriate. On a hillside development, for example, a one-to-one slope (45 degrees) was considered safe for cut slopes and even for some fill slopes; a one-and-a-half-to-one (horizontal to vertical) slope was considered safe for a typical filled slope. Even though many of us knew that these were things that should be avoided, those minimal requirements were about all there was in the way of standards to be met.

Scott: And you knew that those requirements were really not adequate?

Crandall: Oh, yes, obviously. But developers were able to find people who would do the grading—just in conformance with the minimum requirements at the time—which was not adequate.

Also, the drainage characteristics of soils on development sites were not controlled, or did not follow any engineering requirements. The compaction was considered not too important, and houses were built on poorly compacted soil. There were all sorts of problems. Many, many things were done very, very badly.

Scott: That was one of the first steps in improving the city's hillside code regulations?

Crandall: Yes. The next step was to say that fills give us trouble, so we will not make them steeper than one-and-one-half to one, a flatter angle. That went along for a while. We had some heavy rains in Los Angeles in the early 1950s, and there was a great deal of trouble, much settlement of fills and failures of the fill slopes.[4] But widespread hillside development had occurred right after World War II. With

---

4. Rainfall in downtown Los Angeles over the winter of 1951-1952 was 26 inches (660 mm), or about 1.75 times the long-term average. There had not been a season of such heavy rainfall since the winter of 1940-1941, which had a rainfall total 2.2 times the average. There was little hillside development in Los Angeles as of 1941, and what there was tended to involve small-scale cut-and-fill grading on individual lots. By 1951, many massive cut-and-fill projects had been accomplished for tract housing that was developed during the economic boom after World War II.

those rains, it became very apparent that some-thing needed to be done. There was a public outcry, of course.

So a group of us were called in by the city, and we tried to come up with a grading ordinance that would minimize these kinds of failures. I was one of those selected to work with the city building and safety people, to put some teeth into the policy and come up with a sensible ordinance that would minimize these prob-lems. One of the things we required was that a soil engineering firm be retained to do certain things. They had to make an investigation and a report before the fact—before they started doing grading work on the property—stating what were safe angles for the cut slopes and for the fill slopes. That report was then submitted to the building department, and they would review it. They would review the grading plans, and they required in addition that a civil engineering firm prepare grading plans show-ing what was to be done and where.

Much of the previous development was done almost on the back of an envelope. A developer would say, "I'm going to cut this and fill that. We'll do it in this manner." Then they got out there on the site and they did almost anything they wanted. There was no specific set of plans.

So the new Los Angeles grading ordinance required the grading plans and the soil report before the permit was granted. Then, during construction the requirement was that the soil engineering firm be present and make sure that the site was properly prepared, which included removing any topsoil or brush before they started putting fill on top of it. In the past, it was not unknown for the developer and

his earth-mover to just go ahead and place fill on an unprepared site. That is the cheap way to do it, no question about it. It was also the source of say 90 percent of the problems with the fill—the fact that they had not cleaned out the loose material below. Another factor was they had not provided drainage capability, so with the natural drainage blocked by fill, water just builds up in the fill, causes hydrostatic pressures, and weakens the soil. This led to a great many of the failures.

Things like that were covered by the 1952 grading ordinance of the City of Los Angeles, which was way ahead of its time. In fact, Los Angeles was the first area that required this. Other agencies and governments followed suit shortly after, particularly the County of Los Angeles. So that brought the soil engineer into the act.

Failures on Dipping Beds

Crandall: Then in the late 1950s or early 1960s, heavy rains caused problems with excavated slopes in the Santa Monica Moun-tains, particularly the north side. The beds dipped to the north, and the north side of the Santa Monica Mountains was where they were cutting into those dipping beds. That resulted in what is called daylighted bedding—in other words, the slope intersects the bedding planes, and the cuts then had nothing buttressing them, as they did before. Residential proper-ties in the mountains have failed because the soil engineers who worked with the develop-ers hadn't recognized the inclined bedding as a weakness that needed to be considered in analyzing hillside stability. Most of the soil engineering firms assumed that bedrock was just what the term implies, an unyielding formation that wouldn't create any problems. It's bedrock, so what could go wrong?

Scott: But that's not necessarily so.

Crandall: Definitely not, especially when the stratified bedrock is tilted, and some of the layers of it contain bentonite clay, which is very slick when it gets wet. If water gets in bentonite, it becomes like grease. It is like a deck of cards that you tilt and the cards start sliding. So the engineering geologists came to the forefront. They were a very politically astute group, believe it or not, so they managed to pound their drum very hard.

So the code was tightened again, and this time it included a requirement that a report by an engineering geologist cover the bedrock conditions. The City of Los Angeles came up with the requirement that hillside properties require a report by an engineering geologist as well as a soil engineer, before the city would grant a permit to build.

It seems ridiculous that could have happened. Any soil engineer worth his salt—which I think we were—recognized that issue and considered it in tract work. We avoided tract work for several reasons that I will enumerate. So Crandall and Associates weren't directly involved in tracts in these areas. But those tracts were the bulk of the work of other soil firms. Many of them were doing it on a slam-bang basis, where the compacted fills, the excavations, and so on, were done without the kind of thought that the work really should have had.

Scott: That major change probably also

opened up some new territory for soil engineers. Did it result in a much wider realization that soil engineers are needed?

Crandall: It definitely did. It stimulated the utilization of soil engineering firms. There are dozens of them now, of course.

Claims and Litigation, Tract Housing

Crandall: The part of it that really hurt the soil engineering people was that for many of these developments a corporation was set up to develop the project, and then after the project was finished, the corporation was dissolved. So the parties who were responsible for all these decisions regarding quality of the work were no longer available legally. So who is left? The poor little soil engineering firm was still around, because they were a continuing business, and they became the pigeon of the legal profession, and were being sued generally. So if anything went wrong, the only one really left was the soil engineer, or also occasionally the grading contractor, though they managed to avoid the exposure by saying, "We did what the soil engineer told us to do." Many soil firms really got hurt in litigation, for matters that were actually not their fault. They were an unfortunate participant, but had no say over all the criteria.

Nowadays there still can be problems, of course, and the soil engineer has an exposure, but the magnitude and the frequency of problems are much less, so it's a livable situation. Some firms, however, like my own, avoided hillside tract work completely. Even now, LeRoy Crandall and Associates, which became

Law/Crandall, does not solicit tract work. Not that tract work is beneath our dignity, but the original stigma is still there.

A second concern is the fact that the soil engineer often will not get sufficient funds to do the thorough type of investigation that my firm insists upon for a tract. If you did get a developer who promised adequate funds, frequently you were not eventually paid the full amount. Many of the developers were rather shaky, financially.

The third factor, and the key to much of this, is the inspection work. You have to have trained people watching what is done, to make sure it is done properly.

Scott: Are you suggesting that inspection of tract work is much less careful or adequate than it is for other types of developments?

Crandall: Essentially, yes. The developer is working at various different places, and you will very seldom get enough inspectors, paid for on-the-job inspection, to watch every piece of earth-moving equipment.

The fourth and final point with regard to our avoiding tracts is that homeowners, with a life investment in a home, are not going to lose a lawsuit, even if the soil engineer is not at fault. The homeowners win, and the lawsuit and judgment includes anybody who is around, and is financially able to participate in the judgment. Fortunately, most soil engineers have liability insurance. But they are vulnerable to these kinds of problems.

So it looked like there was a better field for our services in major construction, rather than tract housing work; but tract housing remains a big field for soil engineers.

The stimulus to soil engineering provided by the grading ordinance wasn't limited to tracts, but affected any kind of building where it could be shown by a site-specific study that the soil had better characteristics than the building code assumed. Also, on the other side of that coin, the code is not always conservative enough. There may be a condition or special case where, if the code minimums are followed, there will be problems with the building.

In either case, you need to know about site conditions. The problem of expansive soils, for example, is very acute, but it doesn't get headlines. That is because an affected building doesn't collapse, it just gets all cracked up and becomes very difficult to live in. It is quite expensive to repair that kind of damage. We were able to convince people that we could provide factual information about their projects, and that whether what they learned was good or bad, in either case they needed to know. Fortunately, most of the time we were able to show that savings in construction costs based on the findings of soil investigation more than offset the costs of the investigation. That's a happy position to be in, of course, and that's what made soil engineering not only popular, but almost a necessity.

Scott: It achieved that level of acceptance in the postwar period, say up to about 1955. Is that more or less what happened?

Crandall: I think that fits very well.

Development of the Tie-Back Anchor in Southern California

Crandall: I want to mention the tie-back anchor, which was also essentially a southern California development.[5] It is a tie-back for shoring excavations. The model grading ordinance of the City of Los Angeles and the tie-back anchor are widely used foundation and soil engineering approaches that were pioneered in southern California. Probably somewhere in the 1960s, a man here named Joe Lipow developed a machine for drilling a slanted hole into the face of a vertical excavation. The hole typically would angle downward, rather than be straight horizontal. He drilled the hole, removed the cuttings, and formed a bell (an enlargement) at the end of the hole. The hole itself might be thirty feet or forty feet long.

Scott: The bell was an enlargement at the deep end?

Crandall: Yes, the tie-back was roughly a 6-inch diameter hole drilled to about 40 feet in depth, and then a little diamond-shaped belling bucket was used to enlarge the end of the hole. A steel rod was placed in the hole, and the hole filled with grout (essentially concrete) to form an anchor in the ground, a dead-man type anchor. That's the terminology that was used then. The anchor was attached to the face of the vertical excavation, with something like an oversized nut and washer. It held itself by the bootstraps, with the soil anchor providing

a lateral resistance to the face of the vertical excavation. The vertical face of the excavation was tied back deeply into the soil or rock, and it was called a tie-back. Thus there was no interior bracing in the excavation.

The tie-back anchor was a marvelous breakthrough in construction. Prior to that, holes in the ground for the basements of buildings or subterranean parking levels were supported by installing struts across an excavation, running completely from one side to the other. By strutting across, one side pushed against the other. The strutting is not practical with a very wide hole, of course, in which case they would put in what are called rakers—braces angled from the bottom up to the side of the excavation. It is a very tough job for the contractor to work around all those impediments within the hole. By engaging the soil beyond the excavation, and using that as the method of restraining the excavation face, there is a completely open hole, allowing the contractor to work almost as though he were right on the surface of the ground. This technique was pioneered here, and I'm happy to say that my company and I were intimately involved in developing the method, and used the system on some very deep excavations. Now, in southern California anyway, almost every site uses the earth anchor tie-back system to allow economical construction of subterranean projects.

Scott: For the record, who was Joe Lipow? Did he have a connection with Dames and Moore?

Crandall: No. He was a contractor who came up with the idea for the drilling equipment. There have been great advances in

---

5. LeRoy Crandall received the Martin S. Kapp Foundation Engineering Award in 1982 from the American Society of Civil Engineers for his work in developing the tie-back anchor.

the system since, but he had the first drill in this part of the world that was designed to do just this job. He got together with a Mr. Webb and got a patent on their equipment. They called it the Webb-Lipow system and promoted it. Joe was the first installer of this type of design. Subsequently, other shoring contractors developed equipment that was actually even better. But the basic idea was exactly the same.

Chapter 7

Engineering Geology and Geotechnical Engineering

In the 1950s, civil engineers who specialized in soils began to work as a team with geologists.

Crandall: When the City of Los Angeles passed its grading ordinance and subsequent regulations that required a geologist's input, it changed the relationship of the disciplines. The geologist is concerned with rock in place, or nature's formations in place, whereas the soil engineer can take the natural material and rework it and come up with another material. The province of the soil engineer extends to dealing with compaction of fill, for example, and doing the analytical work. The geologist tends to study the site by looking: looking at geology maps and looking at the geology in the field as it is or as it is exposed with an excavation. The soil engineer tends to work by measuring soil properties with instruments, in the field or back in the laboratory. When I entered the field, we had the basic tests—static tests of shear strength, consolidation, permeability, and of course moisture and density. Those are still the fundamental tests. There have been refinements in equipment and apparatus. You now have dynamic testing capability. You can test larger specimens and read the results more accurately. You can put them through cycles of loading in various formats to approximate what you think the actual conditions might be. Obviously, we did not have those kinds of refinements in the early days. But we did have the basic types of tests.

A geologist will say to the soil engineer, "You have the bedrock and it's dipping in a certain fashion. You figure out whether it's safe or not, but I've given you these parameters here: that this geologic material is of this type and could behave in such and such a way under certain circumstances." In the 1950s, civil engineers who specialized in soils began to work as a team with geologists.

Registration of Engineering Geologists by the City of Los Angeles

Crandall: At that time engineering geologists weren't registered by the state, or by anybody else for that matter, so the City of Los Angeles set up its own qualifications board. Geologists who specialize in petroleum exploration, or developing water wells, for example, are experts in other areas of geology and don't have the expertise to advise on slopes and other conditions for a construction project, but there was no distinction in the licensing. The City of Los Angeles building department deserves tremendous credit for all of this. They saw the problem, and they went out and brought in experts to help them come up with ways

and means of providing safeguards against the practice of what was, in effect, engineering geology by geologists who weren't qualified to do that.

So an Engineering Geologists Qualifications Board was set up in Los Angeles in 1957. I was a member of that. There was another soil engineer on the board and then about two or three engineering geologists of note, including Dr. Thomas Clements of the University of Southern California (USC) and Richard Jahns of Caltech, who later moved to Stanford. I think later Jim Slosson was another. We would give an oral interview to the candidates who wanted to be qualified by the City of Los Angeles in order for their reports to be accepted by the building department.

So we would interview the candidates who wanted to get this qualification. I don't know what percentage got through, but probably we felt about 50 percent who applied were capable and 50 percent were not. One of the requirements that we made was that they must be familiar with the geology of this area, not just be a paleontologist, for example, who knew all about the bones of dinosaurs and formations they were found in, but who might not be familiar with what the geology in the City of Los Angeles was.

Scott: That probably meant having previously practiced in the area.

Crandall: Either that, or really having studied and read a lot of literature about the formations, what they are, how they behave, and that sort of thing. This would be a practicing person, who would go out on a job and map the area and come up with the geologic information that the soil engineer and the civil engineer needed to make a suitable grading plan and make sure that the cuts and fills were safe.

Scott: From what you're saying, I take it an engineering geologist is more of a geologist than an engineer. Where does the engineering come in?

Crandall: Engineering geologists are geologists who specialize in construction-type problems. They have to be familiar with what the engineers need to know about the formations and how they're going to function.

When the City of Los Angeles began to require and license engineering geology, that was a real impetus for the whole field. Those universities that only had general geology began to think in terms of that specialty. Schools that didn't have engineering geology began to put in the specialty, because of the desire for it on the part of their students. Once the City of Los Angeles got started, the County of Los Angeles started doing the same thing, in 1959. It got to the point where there were too doggone many governmental agencies with their finger in that pie. If you were qualified by the City of Los Angeles, the County of Los Angeles said, "That's fine, but if you're going to work outside the city limits in the jurisdiction of the county, we want you to talk to our own qualifications board and be licensed by us. It got so that to work in southern California an engineering geologist would have to go through four or five exams.

State Registration and Certification

Crandall: Civil engineers already had a state licensing category, but there was a need to have a better way to regulate geologists. In 1968, the state created a registration for geologists with a specialty certification for those geologists qualified to do engineering geology. I have to be careful about terminology. There are "registered geologists," and there are "certified engineering geologists." You can be a registered geologist in the State of California, which is the broad area of practice, but if you want to be accredited as an engineering geologist, then you have to get certified in addition to being registered. Prior to the 1952 Los Angeles grading ordinance, engineering geology was a name-it-yourself specialty in the geological consulting field. Then it became a restricted field, first by local ordinances in Los Angeles and other local governments in southern California, and eventually, as I'll explain, by state law.

Things have changed more recently, but soil engineering was originally only regulated as a profession by the basic civil engineering license. It was left up to the individual civil engineer to exercise restraint on their own part as to whether or not they had the expertise to practice as a soil engineer. The only discipline within civil engineering that required supplemental registration was structural engineering, and that was required in order to qualify to design California public school buildings, and later, California hospitals. But a civil engineer—like I am—theoretically can practice in any area of civil engineering. I can do structural engineering, as long as it's not a school or a hospital, except that the law says, "You must be experienced or qualified in that area to do that specialty or any specialty." It would be ridiculous for me to try to design a major building, or even a not-so-major building, because I've not done any of that. I did that in school and got exposed to it, and I think I know enough about it to realize that I'm not that good at it. I suppose I could design a building, but it would take me many times as long as somebody who does it all the time, because I'd have to go back to the books and carve my way through in order to do anything.

Geotechnical Engineering Recognized

Scott: Talk a little bit about geotechnical engineering as a discipline.

Crandall: The civil engineering specialty devoted to soils and foundations was first called "soil mechanics." Then it became "soil and foundation engineering," but that got to be a little much of a mouthful. Besides, a lot of people were doing things other than just foundations, so the term "geotechnical engineer" was coined. That is kind of redundant. It really should be geotechnician, but everybody wanted it to sound more impressive than that.

Scott: Geotechnical engineer does sound better than geotechnician.

Crandall: Right. It came about through the American Society of Civil Engineers, and while I had a role in that, I didn't make all these things happen. I just happened to be there. I was the liaison national director to the geotechnical division, or what at that time was called the soil and foundation engineering division, of ASCE. The pressure was on to come up with a standard term for the discipline that was shorter than "soil and foundations engineer." We called for suggestions from the

members and people sent them in. "Geotechnical engineer" was the most popular and most often submitted term. The executive committee of that ASCE division recommended changing the name to "geotechnical division." This was approved by the ASCE national board of direction.

Geotechnical Engineering and Structural Engineering

Crandall: The work of the geotechnical engineer is not as prominent as that of the structural engineer. I sometimes say that the doorknobs get more consideration than the foundations of a building. Once they are built and in the ground, nobody knows or cares about the foundations, unless something goes wrong.

With regard to seismic considerations, however, we come back to the fundamental phenomenon—the shaking a building undergoes comes from the ground. It is the ground movement that affects the building, and until you know what the nature of the ground movement could be, you can't really make a credible design for a building. Now, prior to geotechnical engineering reaching an advanced state of the art, that ground movement was estimated using somebody's guess, or it might even have been the 1940 El Centro record, which received so much usage in those days. We can talk more later about the collection of many more strong motion records in recent years.

And now, geotechnical engineering has progressed to where we can come up with really supportable data on ground motion for the geological environment of a site and the seismic characteristics of that area. In that sense, geotechnical engineering is now a very fundamental part of the seismic design of a building. It provides the needed information that permits a realistic appraisal of the building's behavior in an earthquake.

In the earlier days, we talked about geotechnical engineering merely giving some broad general statements about the type of shaking that might occur. In other words, will the shaking have a high frequency or low frequency? Is the amplitude likely to be large or small for a given site? We could usually deduce this, whether it was a hard soil or a soft soil, since a large-amplitude and a low frequency go together on soft soil, whereas on rock you generally get a smaller amplitude and a higher frequency. In earlier days, that was about as far as we could go.

Site Selection

Crandall: I think most engineers have the feeling that we can build on almost anything if you will give us enough money to do it. Though I guess there is a practical limit on many sites—where the cost of foundation construction is more than the economic value the completed structure would warrant. In that sense, there may be sites that one would say are not buildable, but I like to think that it is an economic determination based upon a proper soil evaluation.

You might come up with unbuildable sites in a residential hillside area where a landslide or potential landslide involves a multi-lot problem. If you are considering building on one of those lots by yourself, it would cost you so much to abate the problem singly that it would just be outrageously expensive. In those

cases, what one does is try to get all the parties together and make a complete fix that is economical, with costs distributed say over six or seven lots.

Scott: That is probably where the assessment district idea would come in.

Crandall: Exactly. But I have been successful in about three out of a multitude of cases of multi-property problems, trying to repair them. There is always somebody who doesn't go along with it or refuses to carry his or her share of the burden. So I consider it a tremendous success to get a group of people together to finance a repair that affects all of them. And it affects not only the people involved, it also affects the surrounding area. Any area with landsliding problems has an effect on adjoining properties. When this sort of thing happens to a neighborhood, the value of the property goes down, and the ability to sell an individual property goes down.

Scott: Are you talking primarily about situations where they are trying to retrofit in an existing development that got into trouble? Or are you talking about new territory that they want to develop—a new development?

Crandall: The problem usually involves existing developments, where the properties are owned by individuals. Having been built on, that makes it more complicated to repair. With new development, of course, you have the opportunity to identify the potential landslide areas and make the fix at that time. If the corrective work is economically excessive, then the developer has to make a decision. Does the developer abandon the whole thing, or set aside those lots as space for a park or something of that sort? There are areas in southern California where there are a great many existing landslides in the natural terrain, and if they are to be developed that must be corrected.

Some of the ground failure susceptibilities are not apparent until you go in and do the exploratory work. That is one reason why geotechnical and geological studies are so important in hillside areas. You can't always tell by looking that an area is landslide-prone, or has had past landslides.

Scott: Especially, I suppose, when they are old or ancient landslides.

Crandall: That is right. A lot of problems have occurred just because an ancient landslide was not identified. It may have been essentially stable under natural conditions. But when you put in a series of home sites, you may change the slopes, and you introduce lots of water to irrigate gardens. The irrigation can be more critical than the natural rainfall in southern California for slope stability. You introduce water into those soils and activate the old slide.

Portuguese Bend

Scott: I guess the Palos Verdes situation on the southern California coast is something like that.

Crandall: Portuguese Bend was a known ancient landslide—it was on geologic maps. In my opinion it probably would have been a successful development, except that it was an area where they used independent sewage disposal systems. They had cesspools and septic tanks.

Scott: That was about the worst possible thing they could have done.

Crandall: Exactly. It was. Not everybody agrees with me, depending upon whose ox is being gored here, but I feel that it was the introduction of water into these bentonitic clay seams, which are very thin, like an inch thick—I think that precipitated the slippage. And the slippage is still going on.

Los Angeles County got stuck with fixing that, which was a miscarriage of justice, in my opinion. Well, it hasn't been fixed, but the county paid off those people. Some of them are still living there. An interesting sidelight on this was that the land movement was so great that there had to be an agreement regarding the property lines, the survey lines. There had to be agreement that the property lines would move with the ground. Otherwise, you might find your house on your downhill neighbor's piece of property. A survey, of course, is based on fixed points in space, unless you have legislation saying that the survey points are moving.

This was worked out, and it was good except for the poor owners who were down at the bottom. Their property ended up in the ocean. Do they have to go to the back of the line at the top of the hill, which is now a gigantic fissure? I am not sure if that was done, but I do know that it was agreed that the property lines moved with your house. That was quite a serious legal problem for a while.

Environmental Contamination

Crandall: Say you estimate your costs based on the information furnished, and then you start construction and run into something different. That can create a problem. One thing that is now affecting project cost is contaminated soil. It really is a hard thing to predict.

In Los Angeles, they used to have a gas station on every corner. Now it is a savings and loan, and the gas stations seem almost nonexistent. But those gasoline storage tanks would leak over the years and go undetected unless the operator began to lose too much gas and tried to find out what was wrong. But practically all of them leaked, and there was disposal of drained motor oil, which was dropped in a hole or a pit in the ground. These things all penetrated the soil and many of them reached or eventually could reach the water table.

We are now aware that this is not a good thing. In fact, I think the pendulum has now swung too far, to the point where we are getting ridiculously concerned about some of the conditions. But in any event, let's say you are an owner or builder and you acquire a site, and have not given any thought to the possibility of contaminated material on the site. You start your site work, say excavating for a basement, which is typical, and run into old seepage in the soil from a gasoline or oil storage tank.

In the old days that was considered not too bad, and people would use that fill and put it out in the parking lot and use it and not give another thought to it. The gasoline eventually evaporated, you know. But now, it has become a toxic waste problem. You used to be able to take the soil in your excavation and dispose of it on a landfill for maybe $1 or $2 per cubic yard, or sell it to somebody else.

Well, that does not work now. That kind of soil is considered toxic, and you can only dispose of it at certain types of landfills, and if you have any kind of PCB material or other toxic characteristics of that sort, which is common, you have to haul it up from Los Angeles to San Luis Obispo County or somewhere like that, a few hundred miles away, and it will cost $100 to $200 per cubic yard.

Costs like that can kill a project, of course. And there is the litigation as to who is at fault and who should pay. The EPA and the county health agencies and other people don't care about cost, you just have to get it out of there and take it away. So contaminated soil is a real problem.

Nowadays, a soil engineer has to consider soil contamination, and do a lot of research on the past history of a piece of property to try to find out if there was any possible source. Environmental assessment is a new thing that is done almost religiously on any site in an urban area to try to find out its past usage. This is done even before a site is purchased. And woe befalls the soil engineer who misses something that did exist before, and he was unable to find it. It immediately becomes your fault—"you should have drilled more borings, you should have put a boring here, you should have checked these records," or something else. The impact is really tremendous.

We had one site in Marina del Rey, where it was known that all of these things had happened—there were underground tanks, there had been welding and machine shops, and manufacturing. They would have acid pickling baths for the steel that they used, and dumped it out on the ground or in the pit. The environmental assessment study found these things out, and now the developer has to clean that site up. This meant millions of dollars. It was a big site.

We used bacteria on that job that will clean up the soil. They have to dig up the soil, spread the soil out, add a food for the bacteria that is sprayed into the soil, and these little rascals go to work and in about a couple of weeks the hydrocarbons, which is what the contaminant is, are gone or greatly reduced.

Scott: That fast? I guess they digest the hydrocarbons.

Crandall: Yes. And you turn the soil over and mix it up bit. The next worry is, "We've created these bacteria, now are they going to take over the world?" Apparently, however, once the food source is gone, they disappear. They die or go into hibernation or something. This method is not cheap, but it is a heck of a lot cheaper than in this case paying close to $1,000 a yard to haul the soil away. It is probably costing about $100 a yard to give it the biological treatment.

In addition, the groundwater on the site is contaminated, as is the groundwater of the whole surrounding area. You as owner of the site are not only responsible for cleaning it up, but if you contribute any polluted groundwater to an adjacent site, then you are also responsible for that. So what do we do? We put a bunch of monitoring wells around the periphery of the property. If something comes into this site from off-site, which is likely, because the ground water is slowly moving towards the harbor, you can say, "This came from that

guy off-site, I did not contribute it to the water supply." Not only is the clean-up required, but also a monitoring program that they will have to maintain for the life of the project.

Scott: Speaking as a lay person, the groundwater contamination seems to many of us more scary than many of the other kinds of site contamination. Groundwater is a precious resource, and if contaminated, the contaminants do not stay in one place, but travel.

Crandall: Contamination has ruined many wells in southern California, in the San Gabriel Valley and San Fernando Valley. They have now set acceptable limits for contamination so doggone low that it is almost impossible to get by. We have to treat wastewater almost to a quality better than drinking water in Los Angeles. There are a lot of problems there. There are no "absolutes"—you are dependent on each individual agency and individual inspectors in the agency, who kind of call the shots on jobs in their territory. You never know exactly what is expected of you.

Asphaltic Sands at La Brea

Crandall: I had one experience with a site at the La Brea Tar Pit, which has been there since prehistoric times.

Scott: Since the time of the saber-toothed tiger.

Crandall: Yes. It is asphaltic sand. The sand contains the asphalt that comes up in the La Brea Tar Pit and causes a methane gas problem in that area. It used to be considered very good fill. You could almost excavate it, mix it with sand, put it in a parking lot, and roll it to produce paving.

Now, it is considered hazardous. We had one heck of a time talking them out of forcing the developer to take that soil clear up to a special dump at Casmalia, a couple of hundred miles up the coast, to get rid of it. This particular inspector from one of the city agencies made the statement that asphaltic sand was toxic, and not only that but also all the asphalt paving in the City of Los Angeles was toxic. If this guy had his way he would have all that asphalt paving taken up and disposed of some way. That is how far some of them can go. I think the pendulum has gone too far.

Liquefaction, Settlement, Landslides, Soil Compaction

Crandall: Microzonation also concerns liquefaction, landslides, settlement, and soil compaction—not just ground shaking. Soil does not necessarily have to liquefy, it can settle without liquefaction if water is not present, and that causes damage to buildings and roadways and pipelines and other things that we depend on.

The degree of shaking expected at a site can also be included in microzonation. That way the land use planners and the builders are made aware of the risks in a given area.

Scott: So microzonation really is trying to specify more precisely what to anticipate in an area that is a smaller part of a much larger seismic zone.

Crandall: That's right.

Scott: This discussion helps me understand better what microzonation is all about. It has always sort of baffled me. I have heard people use the term in different ways.

Crandall: It sometimes depends on the context, depends on what your attention is focused on. It might be microzonation of strong earthquake shaking. Or it might be one or more of the ground failure hazards, like liquefaction or landslides. It might be settlement of man-made fills. Most any geotechnical phenomenon or risk can be subject to microzonation.

I think there are excellent microzonation maps of the San Francisco area, showing where the old Bay shore was and where today's buildings are. Also they show the harder rock areas, and the areas that are good from an earthquake standpoint. You are a hell of a lot better off to build on rock than, as the Bible says, to build on sand—particularly in earthquake conditions. When you know that, if you are interested in safety, it is a useful thing to be able to crank into your planning. I never thought of it that way before, but each comprehensive geotechnical report on a site is in effect a microzonation of that specific site. One thing that you have to be careful of is that you don't focus too sharply on only your particular piece of property, because the lot next to it might have an effect on yours. For example on a hillside, if the slide occurs off your site but comes onto your site, you've got a problem even though you yourself did not contribute to that problem.

Anyway, the need for and the availability of these more detailed maps has really grown considerably. I think that following the Loma Prieta earthquake, there will be much more of that kind of thing available for the public to consider.

Scott: I guess the growth in the need for microzonation reflects a greater demand for it, and awareness of its value. Clients are becoming more aware that, if you know more about an area and know it more precisely, the information can help you avoid future damage. Is that basically what drives the demand?

Crandall: Yes. The great advantage, of course, is primarily for new structures. By knowing what hazards there are on a building site and what potentially could occur, like liquefaction, you can design a foundation that can resist that. So you can build a structure even if you are on a poor site. We do that all the time.

Chapter 8

Soil Engineering and Earthquake Engineering

An earthquake is a real, full-scale test of a building, and unfortunately, we still need that type of test to verify and advance our field.

Early Days of Strong Motion Study

Crandall: Bill Moore, and Ralph McLean were among the early pioneers in the program of the U.S. Coast and Geodetic Survey to deploy accelerographs. In fact, the Coast and Geodetic survey got a reading from the El Centro earthquake in 1940, and that reading became the Bible for engineers in studying earthquakes. While knowledgeable engineers felt that it wasn't necessarily typical of every earthquake, it was the best record they had.

Scott: I used to wonder about that when I was first getting into the field of earthquake hazard. Henry Degenkolb and John Blume and others used to refer to that El Centro record as if it were something very important, sort of like the Holy Grail.

Crandall: It was. If it is the only cup around, it is pretty holy, isn't it? I remember John Blume talking about going to the state legislature and asking for funding for more instrumentation. He told them, "We have only one especially useful strong motion record, from the El Centro earthquake in 1940." One of the legislators responded, "Well, if you have one, what do you need any more for?" But all earthquakes are not alike.

Incidentally, there is a reason for using the term "strong" motion. Caltech, Berkeley, and other places in California, have obtained and kept the seismological record on those large drums with paper around them, as have seismological laboratories around the world. The instruments used, the seismographs, are very, very sensitive. They are used for measuring earthquakes that may be very far away, say in Japan, somewhere on the other side of the earth from the recording station. If the earthquake is close by, the seismograph will try to take a reading, but will jump off scale due to its high sensitivity.

What engineers needed was something to tell them the acceleration of the ground and the acceleration of a building caused by a relatively nearby earthquake. So a whole new stable of instruments was developed, called "strong motion" instruments. They would stay on scale. So strong motion instruments—accelerographs—are used in buildings and in free-field installations so that we get a full record of what is happening at those particular locations. The seismographs the seismologists operate are really intended to tell you what the earth science event was at its source, perhaps thousands of miles away.

You don't want the strong motion recorder to be set off by every little mild shake. They have a triggering mechanism that does not start the recording unless the acceleration reaching the instrument is greater than 0.05g. That is the vertical measurement. The vertical waves arrive first, and if they exceed this value, that starts the camera going. The film begins to move and it is recording when the more damaging shear waves arrive. Later, digital instruments were invented and have largely replaced the ones that recorded optically.

San Fernando Earthquake

Crandall: In the 1971 San Fernando earthquake, a strong motion instrument at the abutment of Pacoima Dam gave some very high readings. Several people have made a career out of studying that. They got readings over 1g. Then there was much study and interest in questions such as whether that severe motion was due to a focused earthquake effect over one small locale, and what was the effect of a sharp bedrock ridge that this instrument was mounted on. The earthquake was the boost needed to start up the California statewide strong motion program, the Strong Motion Instrumentation Program (SMIP), and that deserves more discussion later.

Even before the 1971 San Fernando earthquake, most engineers were aware that strong motion records of earthquakes were essential to intelligent building design for seismic resistance. Then the San Fernando earthquake occurred. It was a very frightening event. I experienced it here in the Los Angeles area, on February 9th, 1971. Our home in La Cañada was on pretty solid material, so we did not suffer any damage at all, but we certainly knew it was shaking at 6:00 a.m. that morning. In any event, that earthquake precipitated a greatly increased awareness of the value of instrumentation.

The City of Los Angeles, through its building and safety department, had already adopted a requirement in 1965 that all structures six stories and taller had to have strong motion instruments installed at the base, mid-height, and top. A number of buildings had been instrumented in time to collect records from the 1971 earthquake.

Scott: How was the Los Angeles program funded? Was that a levy on individual buildings?

Crandall: Yes. The individual owners had to do this. They were required not only to provide the instruments, but also to provide suitable locations for the recording instruments. The owner paid the tab for this, and—what was more of a problem for them—they had to provide space in a room or a special location that could be locked off, so the city personnel could inspect the results. Of course this was not too popular with the building owners, but it was done, and the city was able to enforce it.

The ostensible purpose of the ordinance was to provide information that would be useful in evaluating the safety of a building after an earthquake. Some of us felt that having only three instruments was not enough for that purpose, but it sold the program, and that was very important.

We obtained a large number of beautiful records from the San Fernando earthquake that were a godsend to the structural engi-

neering design people.[6] They had a chance to see how the buildings had behaved when actually shaken.

Scott: This was also a demonstration of the kinds of records you could get through instrumentation, and of their potential value.

Crandall: Yes, it was an example of what these instruments could do, and it precipitated really strong efforts to do something more statewide. When you get a few records, you hunger for more. For example, it turned out that only three instruments per building gave useful data, but was not sufficient to provide all the information needed.

Much work was done on the San Fernando earthquake. I had the good fortune to be put in charge of the foundation study portion of the report that was made on the San Fernando earthquake. That work covered not only the ground motion record, but also what happened, and what did not happen, during the quake.

---

6. The total number of strong motion records from the 1971 San Fernando earthquake was 241, of which 57 were obtained from the top levels of buildings. A few instruments were also located on other structures, such as Pacoima Dam, where the most severe acceleration record in that earthquake was recorded. (R.P. Maley and W.K. Cloud, *Strong Motion Accelerograph Records, San Fernando, California Earthquake of February 9, 1971*. National Oceanic and Atmospheric Administration, 1973, p. 346.) Prior to this earthquake, there were only approximately ten strong motion records of great usefulness to earthquake engineers, out of a total of approximately 100 that had been collected since the introduction of accelerographs by the U.S. Coast and Geodetic Survey in 1932.

Scott: For whom was that report done?

Crandall: For the National Oceanographic and Atmospheric Association (NOAA). They appointed a committee to prepare a report on the earthquake. Martin Duke, now deceased, was the chairman.[7] He had a number of subcommittees, one of which was the soil and foundations subcommittee. I had the honor of being the chairman of that and of directing and working with the people who were studying what happened concerning the ground motion and foundation behavior during the earthquake. This was a very important experience, not just for me, but also for everybody who participated—the engineering profession, and others too, including the social scientists, who were very much involved in finding out how people behaved and things of that nature.

Scott: Say something more about the report on that earthquake.

Crandall: Volume I of *San Fernando, California Earthquake of February 9, 1971* deals with effects on building structures, and is in two parts, Part A and Part B. Volume II deals with utilities, transportation, and sociological aspects. Volume III covers geological and geophysical studies

The report was done under a cooperative agreement between NOAA and the Earthquake Engineering Research Institute (EERI), and was published in 1973.

7. Murphy, Leonard, editor, *San Fernando, California Earthquake of February 9, 1971*; three volumes. National Oceanic and Atmospheric Administration, 1973.

Scott: With a 1973 publication date, that report on a February 1971 earthquake was put out pretty fast.

Crandall: Yes. The introduction was by Leonard Murphy, Karl Steinbrugge, and C. Martin Duke. It really was a very fine piece of work on the quake and its effects. The report collects a series of papers by various authors.

One of the important topics related to soils was the damage to dams, especially the Lower San Fernando Dam That was reported on by both H. Bolton (Harry) Seed, and Kenneth Lee. It came very close to a real catastrophe, believe me. That told us a lot about some of these early dams that were built by hydraulic methods.

Scott: It also prompted the state Division of Safety of Dams to take notice.

Crandall: Yes, the state Division was very prominent in investigating the performance of dams in the earthquake, and finding out why and how. Much came out of that. The safety of dams was one of the biggest seismic safety influences from that earthquake, the greatest in terms of potential hazard reduction. The failure of a single dam can cause disastrous losses. I was out there the day after the quake and it is just incredible how close that dam was to the water overtopping it after the embankment failure occurred. It was a matter of a few feet, as the top of the earth fill dam subsided tens of feet, and if the water had ever gone over, it would have eroded through very quickly and wiped the dam out—almost instantaneously.

I felt obliged to point out in my contribution that a great many structures went through the earthquake without suffering damage. It has always impressed me that everybody takes pictures of the damaged buildings, and of the ground rupture, and from that you can get the idea that an entire city has been demolished. Yet actually only relatively few buildings were damaged, particularly with regard to residential wood frame structures. Few had any significant damage, and even those that were astraddle the fault rupture did not completely collapse, although those that were right on the crack were unsalvageable. The fault rose about three to six feet on one side, and there was also some lateral shift. Nobody was killed in those residential buildings.

Moreover, the homes that were not on the fault rupture, and were recently built, came through very well. Some of the concrete floors cracked, and a few masonry walls shifted a little. Some things did happen, of course, particularly to those that did not have adequate foundations or well-built foundations. Some of the older houses that were not anchored shifted off their foundations. It was, of course, already pretty well known that this could happen. Also, there was damage or collapse of old brick chimneys that were not reinforced or braced.

California Strong Motion Instrumentation Program

Crandall: As a result of the San Fernando earthquake, the California legislature passed a law setting up the state's Strong Motion Instrumentation Program, SMIP.[8] Prior to this, there was the City of Los Angeles program, and in 1970 a similar instrumentation requirement for taller buildings was put in the UBC.

I had nothing to do with the establishment of the state's program, and I take no credit for any of this happening. Certainly I was in favor of it, but those engineers who did pursue this, like John Blume, Karl Steinbrugge, and others, did a very fine piece of work in convincing legislators. I also think Senator Alfred Alquist was probably the most supportive and instrumental California state legislator in this, and he realized the importance of coming up with legislation that created a program of instrumentation.

I was lucky enough to be appointed to the initial steering committee, and one of the others on that committee was Martin Duke. I remember particularly that he and I used to ride the plane to Sacramento together, or to wherever the meeting was located. I can tell you later how the conversations we had on one of those plane rides led to the formation of the lifelines group, the Technical Council on Lifeline Earthquake Engineering (TCLEE), in the American Society of Civil Engineers.

Scott: It's interesting how a development along one front in the effort to extend seismic safety can lead to developments in another.

Crandall: Yes, it is. You're right. So this advisory committee of about a dozen people was set up to guide the new SMIP program, to brainstorm what should be done and how.

Scott: That committee reported to CDMG?

Crandall: Yes, to the California Division of Mines and Geology.[9] In my recollection, Harry

---

8.   Chapter 8, Division 2, Public Resources Code, enacted 1971.

9.   In 2002, the common name of the agency was changed to California Geological Survey, though the previous name, California Division of Mines and Geology, is retained in some statutes.

Seed was the first chairman, and if I am correct, this predated the Seismic Safety Commission beginnings.

Scott: Yes, because the Commission did not actually become active until about May of 1975.

Crandall: Gordon Oakeshott, with Mines and Geology, was very active. I worked on a subcommittee with Gordon regarding the desirable locations of instruments to record the free-field information. The intent of the program was, and still is, to have such coverage that no major earthquake in the whole state of California would go unrecorded by some relatively nearby instrument. Prior to this, there was no master plan or strategy. It was simply a building-by-building process—a building permit requirement triggered when a building was six stories or taller. That, of course, meant most of the instruments were being clustered in downtown districts of cities and weren't well distributed.

Scott: So the intent was to blanket the state with strong motion instruments?

Crandall: Right. Our whole intent was to get instruments out to record what happened to the ground, to answer the question, "What is the ground motion during an earthquake?" The committee acted in an advisory capacity and assisted in selecting locations. We set up priorities for the first places to put the instruments, which was not necessarily in the major cities, but at any spot where we thought we would get the maximum amount of information and as soon as possible.

We did not want to miss any quakes. So one group, the seismological group of the committee, would try to estimate which faults were most likely to have an earthquake, and to suggest some kind of priority. Of course, the San Andreas fault and a number of other faults were well known, particularly up north around Eureka. We felt we needed to get instruments out there because those areas were highly seismic. My recollection is that we did not have much money, but were doing a lot of planning, which was essential.

Sometime after the initial committee work, the permanent funding legislation was passed, under which a surcharge was placed on building permits throughout the state. Those cities were excepted that had already adopted a program of instrumenting certain buildings, for example the City of Los Angeles. I think all together there were about fourteen cities that had programs considered adequate to qualify for the exception.

Scott: So when the state program was set up, there were already over a dozen cities in California that actually had some kind of ongoing strong motion program?

Crandall: Yes. Most of them were pretty minimal. Maybe they required one instrument, or they had some good intentions. As I recall, Los Angeles was pretty much the only city that was doing anything really significant.

After a few years, the City of Los Angeles realized that their maintenance costs were really exceeding their ability to look after all the buildings that had had the three instruments installed under their code. A group of us met with the chief of the building department of the City of Los Angeles, Walter Brugger. We approached the city regarding their coming into the state program. They decided that this was prudent for them from an economic standpoint, because of their maintenance costs. The state would maintain the instruments in buildings within the city of Los Angeles that we selected.

That was accomplished. This was one of the few diplomatic highlights of my career, working with the City of Los Angeles Department of Building and Safety to arrange for this marriage, and the relinquishment of their program to the state program. It all worked out very well, and the city council accepted the changeover. In our negotiations with Los Angeles, the city became very interested, and as a matter of fact, even eager. They had concluded that the three instruments per building they had installed were really not giving all the kinds of information they needed, while conversely, the instruments were also recording more data than they were able to use. There were also some real constraints about using the information acquired by the city program as public information. Building owners sometimes put some constraints on that same data. From our standpoint, that limitation was not a very good thing.

Scott: Owners could put on such a restriction under the Los Angeles city program?

Crandall: Well, theoretically the city had access to the data, but whether or not they could divulge it to the rest of the world was one of the uncertainties. It had not been tested in the courts, and I think they were reluctant to try it, as a matter of fact.

In any event, that gave a boost to the state's strong motion program, when Los Angeles joined it, because the Los Angeles contribution in building fees was very significant. The state also inherited a lot of the instrumentation, some

of which was not used, and that was supplemented by other instruments, including free-field installations. Also, there was a better distribution of additional sensors throughout the buildings that were accepted into the state's program. There was latitude in how many sensors and where to install them in the CDMG program, as compared to the requirement of Los Angeles to install bottom-middle-top instrumentation and only in the taller buildings. As of the 1990s, state legislation required all cities to contribute funds.

Interestingly enough, the staff of the state's SMIP contacted most of the cities that had some sort of strong motion program. When asked what they had done with the records they are supposed to have been getting, most of them were unaware that they even had the program. Other cities did not realize they had been exempt, and some of them had been collecting the money, but did not know where to send it. So the action was not unpalatable to them.

It had been twenty years or so since some of these cities started their programs in the late 1960s or early 1970s, and personnel had changed, city administrators and heads of building departments were no longer the same people that were there when all of this began. So, many people did not have any idea what it was all about.

Under the state program, most of the building departments in the state collected a surcharge fee, a rather nominal one, seven cents per thousand dollars, something like that, 0.007 percent. These monies went into the Strong Motion Instrumentation Program fund for providing instruments and maintaining those that were installed.

It was not very long before we realized that the formula limited the number of instruments we could put out, because maintenance costs and the cost of taking care of the instruments was an ongoing expense that kept growing as more instruments were installed, and maintenance began to take a large part of the total fee.

We suddenly woke up and said, "Hey, we cannot just put thousands of instruments out, because we cannot afford to service them." So there was a limit, and we began to work with that in mind.

Role of the Seismic Safety Commission in SMIP

Crandall: The funding issue was brought up by Bruce Bolt when he was on the Seismic Safety Commission. At some point, between the state program's commencement and the end of the 1980s, the Seismic Safety Commission was required by legislation to supervise the state Strong Motion Instrumentation Program (SMIP), although the operation of the program remained under Mines and Geology, and is still there.

Scott: The program's overall Advisory Committee and its subcommittees all were brought under the Seismic Safety Commission?

Crandall: Yes. I am now a little vague as to exactly how that worked, but at some point there was new or revised legislation, and the Strong Motion Instrumentation Program was brought under the aegis of the Seismic Safety Commission, which appoints and is responsible for the Strong Motion Instrumentation Advisory Committee (SMIAC).

If you read the legislation carefully, you see that the committee only has responsibility for advice. The law does not say what happens if Mines and Geology does not want to take the advice. In practice, however, there is an extremely close relationship between the Seismic Safety Commission and the SMIP program, and the Advisory Committee's advice is solicited and followed very religiously.

Thus, the various subcommittees of the Advisory Committee are involved in selecting the actual buildings and structures that are instrumented, and they set priorities for that program. My present association with the strong motion program [1991] is as chairman of the Advisory Committee. I have been on the committee since I became a member of the Seismic Safety Commission, and in fact became chairman of it shortly after I became a Commissioner.

At the outset, the Advisory Committee was chaired by Harry Seed, who was a Commissioner on the Seismic Safety Commission. He was followed by Bruce Bolt, who was also a Commissioner, and then when Bruce became SSC Chairman, he relinquished the post of Advisory Committee chair, and I was appointed as chairman of that committee, and Bruce stayed on as a committee member.

Scott: I recall that by that time, you had been on the Commission about two or three years. You became chairman around 1983 or 1984? Bruce became chairman of the Commission about six months after Dick Jahns died, in December of 1983. I was Commission chair at that time, and Dick was to take over in January 1984. Bruce had commitments, so he could not take over as chair immediately, so I continued to serve as chair for three or four months into 1984, before Bruce became chair. Bruce took over about May or June of 1984.

Crandall: Yes. When Bruce became SSC chairman, he did not want to keep the job as chairman of the Advisory Committee, although he stayed on as a committee member.

Scott: The Advisory Committee chairmanship was a fairly demanding role, was it not? That certainly was my impression when I was a Commission member. Over the years, the Advisory Committee and subcommittees seemed to be a busy bunch.

Crandall: The Advisory Committee meets at least twice a year. Then the executive group—formed by the advisory subcommittee chairs—meets in the interim periods, maybe once or twice a year. There are five subcommittees that really do the work. For example, the building subcommittee involves about ten or twelve people who all get together. We also have other subcommittees on lifelines, data utilization, ground motion, and now a new subcommittee on directed research. The new committee is seeing that the strong motion data we get from earthquakes are being used to further engineering knowledge—and that is a real success.

Scott: I remember always being very impressed when Harry Seed, and then Bruce Bolt, and then later you, reported to the Commission. It always seemed to me that you had a lot going on.

Crandall: Yes. It is a major effort to oversee the Strong Motion Instrumentation Program. In those years, there was probably a budget of seven to ten million dollars a year.

Scott: Along the way, the tax levied on building owners to support the program increased, didn't it?

Crandall: Yes. We got additional funding a couple of years ago.

Scott: After the program's first ten years or so, building owners' fees were increased to almost double the original amount, I believe.

Crandall: I remember Bruce Bolt preparing a study that showed it would be at least twenty years before we could get in all of the instrumentation that we wanted, under the previous fee level.

Scott: Did Bruce do that when he chaired the Advisory Committee? Or was it after you became chairman of the strong motion committee?

Crandall: It started when he was chairman of the Advisory Committee, because I remember him making the study. It showed the money coming in at a certain rate, and what we wanted to spend it on and how fast. That showed that it would be twenty years before we could get to the point where we thought we should be. So we tried to scale the time period back to five years. That's when the additional funding was sought and obtained, thanks, I think, mostly to Senator Alquist for his strong support.

I must say for the record that the really great advances in the Strong Motion Instrumentation Program were made under the chairmanship of Harry Seed and Bruce Bolt. I am sort of a caretaker chairman, I would say, and the important things were done previously.

Scott: I do not think you are justified in minimizing your contribution, although I agree that Harry Seed and Bruce Bolt did a lot of pioneer work because a lot needed to be done for the new program. They were breaking a lot of new ground.

Crandall: We should also give credit to the program's staff. They are doing very fine work. Tony Shakal is program director. Tony's background is in geophysics.

Scott: He is a member of the CDMG staff?

Crandall: Yes, that's right, and he is assigned to SMIP, which, however, is funded separately from regular CDMG activities. It is almost as if he had a completely separate business. They do their own financing, accounting, and all of that.

Scott: The funding comes out of the fees and charges, so it is all kind of a self-contained?

Crandall: Well, we had thought it was. But then, just last year, we found out to our shock that we were subject to overhead levied by the Department of Conservation.

Scott: The good old state bureaucracy and Department of Finance have their own insidious ways of doing things like that!

Crandall: Moreover, CDMG suddenly woke up and realized that they had a little gold mine in the strong motion operation. So now we get tapped with about 20 percent in overhead charges, which are hard to fight. Previously, we thought we were operating completely independently of the rest of the world. But we got stuck with the overhead charge, much as we hated to see the money siphoned off for that purpose.

The talent on the Advisory Committee is the best—incredible people like George Housner, who started our data utilization subcommittee. Bill Joyner from USGS participated. Roy Johnston and Jerve Jones, for example, are also highly interested members. Jones is with a major contracting firm in southern California, and takes time to come to these meetings.

Initially, it seemed that only engineers and geologists were involved. One thing I think I helped was to expand and broaden the membership and outlook to some degree. Thus, Jerve Jones serves on the Advisory Committee now, and two or three others like that. Mary Henderson was one of the early ones on the original committee. She has always had an interest in this work.

Scott: For the record I should mention that Mary Henderson was very active from the beginning of the Senator Alquist legislative committee advisory groups that predate the 1971 San Fernando earthquake. She was a very active member of the advisory group on government organization and performance, along with Louise Giersch, and Bob Rigney, and me. Henry Degenkolb was an observer on that advisory group, representing the engineering advisory group. So Mary Henderson has a history that goes all the way back at least to 1969 in this field.

Then, when the strong motion program was set up, she was considered a logical candidate, with her demonstrated interest in earthquakes and with her local government background. She was a council member from Redwood City in the San Francisco Bay Area.

Crandall: Her appointment came via recommendation of the League of California Cities. I remember now that in our early strong motion committee work, Mary pushed for the idea of putting out an informative pamphlet saying what the program was all about, and explaining what we were doing, to help people understand. It was a little publicity program.

Scott: She was very aware of the importance of bringing the public along and explaining things clearly.

Crandall: Yes. She has been a tremendous asset, and is still involved in our subcommittees.

Extension of Strong Motion Instrumentation Program to Caltrans, Hospitals, and Schools

Crandall: As of now [1991], Caltrans and the hospitals are or will be participating in California's Strong Motion Instrumentation Program. Certainly Caltrans is participating now, whereas the law previously excluded them because they were not contributors.

The law says something to the effect that if you do not contribute to the state's Strong Motion Implementation Program, you do not participate, and that eliminated all of those activities that did not get building permits through local agencies, which was the basis of the funding. State bridges don't pay for local government building permits. But now, since the Loma Prieta earthquake, they are able to do some funding, and that will enable them to instrument a number of their bridges.

It was a shocking thing to find out after the 1989 Loma Prieta earthquake that there were no instruments on the Bay Bridge or the Golden Gate Bridge or most of the other bridges. An exception was the suspension bridge here in San Pedro, over part of Los Angeles harbor. The Vincent Thomas Bridge had been instrumented.

Scott: Why was that bridge instrumented and not the others?

Crandall: It is the most recently built major bridge. In any event, for our Strong Motion Instrumentation Program, we concluded that we needed some bridges instrumented. With the viaduct collapse in Oakland, and the finding that the design of many of those structures have defects that show up under strong shaking, Caltrans has received a windfall in funding to enable them to study these structures and come up with retrofit work. Jim Gates has managed that program at Caltrans, has been on our Advisory Committee, and is active as a subcommittee chairman. Jim has been pushing for instrumenting bridges and typical structures in the Caltrans system.

We now have the pleasure of knowing that Caltrans is able to participate in our program. Previously, the law prevented them from participating in SMIP, or rather SMIP was prevented by law from instrumenting structures where we did not receive a contribution from the fees on local government building permits.

We were not able to instrument structures under the purview of non-SMIP-funding agencies unless the Commission itself adopted a special resolution of urgency, saying that this instrumentation was important. The last exception granted was for the South Tower of the Golden Gate Bridge. The Golden Gate Bridge District, which controls that structure, is planning to completely instrument the bridge in the future.

But that's a long and very expensive process. It takes a while to develop.

So it was decided that, meanwhile, some information should be obtained in the event of a nearby earthquake by putting out what we call temporary instruments on the South Tower of the Golden Gate Bridge. We could do that in a hurry, so in case an earthquake occurred, that information would be available. That was an urgency case that the Seismic Safety Commission approved, and the instrumentation is now, as of 1991, being installed on the Golden Gate Bridge.

Scott: Does the Strong Motion Instrumentation Advisory Committee now play an advisory role to Caltrans?

Crandall: Yes. Caltrans will submit a group of bridges that they feel are typical, and that they believe would be worthy of instrumenting.

Scott: So they in effect say, "We think these are the right bridges?"

Crandall: Yes. One of the things we don't like to do is instrument a unique structure that will never be duplicated. While it may be nice to know what it is doing, the information does not help in the design of future bridges and structures.

Instead, you want data you can apply to other structures. So the intent is to get to more typical bridges. The information obtained will be useful for designing bridges of the same type in the future. Anyway, Caltrans will submit a number of structures, our committee will screen them, assign a priority, and instruct our SMIP people what to do in the way of installing instruments.

The hospitals in California, because they go through a state permit process rather than local government, were also not part of SMIP, but will now be able to participate through some funding out of what they call a research fund from their permit checking fees. We also have the opportunity for some of the public schools to participate now. Because in California the public schools also go through the state building permit process, rather than local government, they were not subject to the SMIP tax and weren't in the program.

Of course, seismic regulations for schools and hospitals have a long history in California. Public schools were brought under statewide earthquake design provisions back in 1933, after the Long Beach earthquake. The state architect's office then functioned as the building department for the schools, rather than the local jurisdictions. Then the San Fernando earthquake prompted the passage of the Hospital Seismic Safety Act of 1972, which also made the state, in effect, the building department for permits for those kinds of facilities, though it was a different state agency, the Office of Statewide Health Planning and Development. Yet here we were, ignoring getting strong motion data on hospitals and schools, and they weren't included in the Strong Motion Instrumentation Program run by CDMG.

As a committee member on the Strong Motion Instrumentation Advisory Committee, I have mixed emotions about an earthquake. Nobody wants to see a major earthquake happen, but yet we would like to get the valuable kinds of information that will make this investment in instrumentation really pay off.

Scott: That is true, even though so far only relatively small earthquakes have occurred in instrumented areas.

Crandall: Well, there was the 1987 Whittier Narrows earthquake. We had some instrumented buildings in the vicinity, and also got information about high-rise structures in downtown Los Angeles, where the instruments measured the effects. Earlier, in 1979, the El Centro public service building, the one whose failure attracted so much attention, was instrumented. That gave us much information about what happened in the building during that earthquake, which was a rather modest one.

Recording the Way Earthquake Waves Travel

Crandall: With modern improvements, these strong motion instruments are all interconnected now, and are all set to the same time. Data can also be fed back to the collecting station by radio or over phone lines, because it's digital now. So now if, for example, the quake is first recorded at 1:15.00 at one station and you read it several miles away at 1:15.10 seconds, you know it actually took the motion ten seconds to get from point A to point B.

Scott: You can read how the motion travels, and learn how it may change over distance?

Crandall: That is right. And that gives us very, very valuable information when the instruments are in arrays. They string up a set of instruments, it may be for miles, starting perhaps close to a known fault. Then a quarter of a mile away, half a mile, three quarters, five

miles and farther, other instruments will be set at each of those points.

By having the clock times all the same, they can determine how long it took for the initial shock to get from one station to the next, and the next, and so on, and determine how fast the wave is traveling through the soil. They can get that and other information enabling us to predict what might happen at a site that is maybe ten miles away from the source of the earthquake. The attenuation effect is what we are looking for. A number of those arrays have been set out that will provide this sort of information in California.

There is a very famous array in Taiwan, SMART-1, or the Lotung Array, which Professor Bruce Bolt has been very much involved in. It is in a series of strong motion instruments set in concentric circles. They are getting some superb information from that.

Of course, one must remember that each of these locations has its own specific geologic characteristics. While we are getting valuable information from Taiwan, you have to translate that into the kinds of geologic conditions that occur in the area you are interested in. It is sort of like the original [1940] El Centro record. You had one record and you made use of it, but knew full well that your site might not be shaking in exactly that same manner.

Flow of Seismic Forces in Reality and in Building Design

Scott: Let me ask a question or two relating to ground motion and building response. You have several times explained what seems the commonsense interpretation—that is, the forces enter the building from the ground up. But I am aware that structural engineers tend to talk about designing from the top down. I have never fully understood why they talk that way, since common sense tells you that forces go the other direction.

Crandall: In the structural field, we think of it in the reverse—that is, you design a building from the top down. But the actual effect is that the ground moves, and the building, due to its inertia, tries to resist the movement—good old Newton's Law being what it is. Then the ground shaking causes the building to vibrate. Seismic forces do start at the bottom, yes. The structural engineers design by calculating an overall seismic load or base shear, but then they start at the roof and calculate how much of the total building lateral force it represents, then take that load and transfer it to the story beneath, which adds its own lateral load, and that cumulative load transfers to the next story down and so on. Tall buildings that vibrate in several modes rather than in a simple back and forth motion are a little more complicated, but basically that is the explanation. In other words, you are designing it from the top down as far as the structural engineer is concerned, with the forces increasing as you move down from the top of the building. The members and connections are designed to adequately carry those lateral forces all the way down from the roof into the foundation.

Scott: Is it easier to think about or to work with that way?

Crandall: Yes. How can you design the bottom story if you don't know what the loads are from the other stories? So you start at the

top and add up the lateral forces as you go down. When you get to the bottom, you have the whole thing. That is the technique for the structural design of a building.

But, in reality, the earthquake shaking transmits through the ground and through the foundation up into the building. That sets the building to oscillating. Depending on the period of the building, you will develop certain additional forces.

One thing I have always tried to tell people is that if the soil is so weak that the full force can't get into the foundation, then the building does not have to resist it. Let me rephrase it. Looking at it as though the building were transmitting the forces to the soil, there is a limit because if the soil could not transmit the forces in the first place, then the building will not be subjected to much greater forces. I want to emphasize that the soil will yield as it tries to force load into the building, and if the soil is plastic enough—say a building is on driven pilings that goes through soft soils into firm soils or even into rock—then the earthquake force comes from the rock through the soft soil to the building, imposing a lateral force on the piling. Now, if those soils are so soft that they cannot transmit that earthquake load into the piling, then you don't have to design the piling as though it would get the full force.

An Unknown: The Seismic Behavior of Basements

Crandall: The force that has to go out cannot exceed what comes in, in my opinion. For lateral resistance of foundations, and also of basement walls, two big unknowns are the dynamic earth pressures against structures and the seismic behavior of basements. There are wide differences of opinion among engineers—soil engineers in particular—as to what kind of increased loadings you will get on the basement walls of a structure due to earthquake.

Many people think that the load is much greater than it is in the static condition. As far as I am concerned, however, I am not that much a believer in a big increase in loads, unless in the case of an extremely large structure, hundreds of feet in dimensions.

My theory is that the subterranean part of the building moves with the ground, primarily, so there is very little additional load generated. If the building is extremely stiff and resistant, however, then you could develop additional pressure. Essentially, though, for most structures, it is my opinion that the basement and the earth around it are moving simultaneously, together, and in tune with each other.

I haven't proof of that, except that I have not yet seen a structure in an earthquake area where the basement has collapsed or failed as a result of the earthquake. You might say, "Well, that could be because basements are always inherently stronger than we give them credit for." On the other hand, if that is so, we should be taking it into account and not just saying that the soil pressure is much greater and using that in the design. I am talking about suggestions to use twice the lateral force, or up to three times, based on some theories of earthquake design for basements, as compared with what you would have under static conditions.

Not only have I not seen evidence that basements have suffered due to increased soil pressure, but also my firm was conducting tie-back anchor test loadings in a basement shoring situation in the Westwood area at the time of the 1971 San Fernando earthquake. It was a forty-foot-deep hole, and we had load cells and strain gauges on these tie-backs that we were testing continuously, twenty-four hours a day. Although our people down in that hole were shaken by the earthquake and were a little concerned, there was no evidence whatsoever of any change in the load condition of the tie-back anchors.

This happened in two cases. Twice we had a test underway at the time of an earthquake. That is rather limited data to base the theory on, but I include this to indicate the differences of opinion among engineers as to the kind of loading an earthquake will generate in the foundation.

Scott: And those are not just minor differences?

Crandall: No, it can add a lot to the cost of construction of a basement if you include this kind of upper-bound earthquake force. So we still have things to learn about the behavior of structures.

Free-Field Motion and Soil-Structure Interaction

Crandall: One important thing learned in the San Fernando earthquake was the value of having an instrument remote from an instrumented structure—called free-field instruments. In other words, you also need a measurement of the ground vibration where it is unaffected by the building itself, and free of any other encumbrance.

Partly as a result of the San Fernando earthquake, it was recognized that the instruments in the basement of a building were affected by the behavior of the structure itself. Since this was a modified ground motion, those of us in geotechnical work in particular said, "We need to know what the ground is doing before it is influenced by a building." With free-field motion records, you could take the ground motion and apply it to any building and feel confident that you knew what was going to be the input source. The ground shakes, the building shakes, and as the building shakes the motion of its foundation interacts with the motion of the soil around it. We geotechnical people consider it very, very important to know what the ground itself is doing.

Scott: So the free-field instrument gives a purer, less "adulterated" type of ground response reading?

Crandall: Exactly. From free-field readings, you know what motion came from the earthquake and arrived at the site, without any influence of the particular building. The earthquake comes into the structure, and the structure starts its own vibrations, depending on its structural characteristics—height, mass, and so on. It feeds those motions back into the soil because the foundation feels that motion of the building and is in contact with the soil, and there is a soil-structure interaction effect. The basement instrumentation reads the combination of those effects.

Initially, when we first had instrumental readings that were only from the basement or lowest level of a building—in addition of course to the upper-floor instruments—we were utilizing the record of the motion at the base of the building as the input, as the source mechanism for the shaking of the building. Things were not turning out the way we thought they should. The feedback of the soil into the structure and vice versa was giving these modified dynamic readings. Then we started the free-field program. In almost every installation now, we have a free-field instrument in some location where we feel it can measure the pure ground motion at that site, without any influence from the structure.

Here is another important point. In addition to surface measurements, we have now reached the point, and the instrumentation has been developed, so that we can place strong motion instruments in deep holes in the ground and leave them there permanently to measure what is happening at depth. In addition to knowing what is happening on the surface, we are able to install these downhole arrays, as they are called. A strong motion instrument is placed in a boring as deep as 500 feet—that's the deepest I'm familiar with—and another one at, say, 200 feet, and one at the surface. From that, you can learn how the wave propagates through the earth and up to the surface.

Theoretically, the 500-foot depth receives the wave first, then the 200-footer, then the one at the surface. When that happens, we will have data on the transmission effect of ground shaking vertically, as well as horizontally. We're very eager to get some of that data from earthquakes. We have some downhole instruments at the Parkfield site, which will help give us an idea of what the effect is when the next earthquake occurs there. USGS has heavily instrumented that area of California in expectation that another magnitude six earthquake will recur there, as has happened relatively regularly in the past. We know that the earthquake wave changes as it passes through the strata in the ground, and as it gets to the surface it sort of breaks into surface waves of different types, with different wave velocities. We do not know precisely whether that's the same motion as is coming through the earth or not. The only thing we have is intuitive, or is based on knowing what happens at depth in tunnels and mines. Miners sense less shaking of the ground at their depth than occurs at the surface, for example.

Predicting Ground Motion at a Site

Crandall: All of this information is, of course, helpful and increases our knowledge and the state-of-the-art in our ability to predict ground motion at any site, based on a possible earthquake event. That is one thing that geotechnical engineers can now provide, and it is part of the service my company provides—the prediction of ground motion at a given site.

What is the effect of the immediate subsurface conditions on the earthquake waves as they pass through the ground? We know that they change their character as they go from rock to a soil condition, and through soil of different types, such as firm soil as compared to soft soil. So we, meaning geotechnical engineers, obtain measurements of the shear wave velocity of these shallow soils that the seismologist does not even consider.

Seismologists are concerned with deep earth structure and the source of the earthquake, where it actually ruptures and releases energy, which may be ten miles deep. But that motion transmits through the rock and eventually reaches the surface. When it gets up near the

surface, it can change its wave form and the content of its motion. We, as geotechnical engineers, try to determine what that motion would be based on these seismic-type measurements that are made.

The structural engineers can then utilize this in their analyses of buildings. Let's say in Los Angeles we consider the San Andreas fault and assume that a big 8.3 magnitude earthquake might occur there, but also critical—and maybe more critical, depending on location—is the Newport-Inglewood fault, on which the maximum earthquake is not expected to exceed about 7 magnitude, or maybe a little more. It is not as large in total energy, but is much closer to downtown Los Angeles and the west side.

So you get entirely different shaking characteristics from the Newport-Inglewood fault than from the San Andreas fault. We attempt to evaluate both for each site that we study and provide the structural engineer with these data. The structural engineer then checks both out and sees in a particular building design which is the more severe, and must take the more severe motion into consideration.

Because of its distance, and the filtering out of the high-frequency waves, an earthquake on the San Andreas will transmit much longer-period waves in Los Angeles, and that will then have greater effects on high-rise buildings—which have long periods. Whereas, with an earthquake nearer by, such as on the Newport-Inglewood fault, the higher frequencies are not as filtered out, and you tend to get a short-period motion with a more serious effect on the smaller buildings—say buildings of three or four stories.

The same thing happened in Mexico City in 1985—very dramatically. Because of the distance of the main quake and the character of soil in some areas of Mexico City, a long-period motion affected certain categories of buildings, such as those in the range of eight to twelve stories, which took a serious beating—pancaked and so on. Whereas the very tall buildings of forty or fifty stories and the short, stiff ones of two and three stories, did remarkably well.

Value of Dynamic Studies

Crandall: The building code permits deviation from its seismic provisions if there is enough supporting data. This has been true in the case of soils studies. In its simplest form, the building code gives bearing values for certain types of soil, and you can design your foundations for those values. But the code also says that if you have an approved soil engineering report that gives other data, you can use that. This has been a big stimulus for soil engineering in that area.

The same thing is true of the dynamic studies, which simulate how much a particular building will shake. You can use values for the shaking-induced forces from the formulas in the code for up to, I think, 160 feet in building height. Up to that height, you can also use dynamic analysis if you want, but the code requires dynamic analysis for buildings over 160 feet, those of irregular shape, and a few other exceptions. So certainly any building over fifteen stories or so in the City of Los Angeles has to have a dynamic analysis. I think as of the early 1990s, the UBC and the Los Angeles code have been similar in that respect.

In California, we can assume a structural engineer is employed for the design of most of the bigger buildings—the ones beyond the scale of an individual house. The structural engineers will make a judgment on the larger or more important projects as to whether a dynamic analysis would be beneficial. In other words, they decide whether the savings from the dynamic analysis is likely to be enough to warrant the cost of doing it. The dynamic analysis often demonstrates that the building need not be as resistant as the simpler code approach assumes, which means less steel, less concrete, less cost. The building code usually is conservative. So if you do these other things in lieu of just using the code-prescribed formulas and values, you can usually produce some savings.

Scott: In short, if the engineer can demonstrate, with the help of a qualified soil engineer or geotechnical firm, that it is legitimate to do so, they can shave the construction costs by not designing as much strength into the building?

Crandall: Yes, that's basically it—using the geotechnical information as input, and then the structural engineering dynamic analyses, which is now done easily on the computer. You need both the geotechnical engineer's input and the structural engineer's dynamic study. I think that in southern California, most any buildings of significance are getting dynamic analyses, because the programs for this sort of thing are available now. Almost any practicing structural office has access to them.

Scott: So it is now really part of the state-of-the-art, and another effect of the advent of the modern computer?

Crandall: Yes.

Existing Buildings: A Dilemma

Crandall: The big problem, of course, are individually owned existing structures. With a new piece of property that a developer is subdividing and getting ready for construction, there is a chance to work with dynamic considerations for soil and design, based on soil engineering and geological factors. The big rub is—what do you do with the buildings that are already built?

Scott: San Francisco's Marina District is a classic case. What can they do with those buildings? Some of them are lovely looking buildings, but if another shaker comes, there is that "jelly" down underneath.

Crandall: The good side of it for the guy who owns one of those buildings is that the public forgets pretty quickly, and in a few years their value will be back up and maybe even higher. I remember in Los Angeles we had the Bel Air fire in 1961, which was a real disaster in a very, very affluent area, burning almost 500 homes. For a while, nobody wanted property in Bel Air because of that. But in two or three years' time you wouldn't have known that anything had happened at all there to the real estate market.

Long Seismic Return Periods

Scott: I would like to ask about one item listed in the program of the 1991 Second International Conference on Recent Advances in Geotechnical Earthquake Engineering and Soil Dynamics. It is "Design for environments with long seismic return periods." I guess that refers to places where earthquakes are expected, but not very often.

Crandall: Right. Maybe every couple of thousand years. The average or mean period of time before the return of the large earthquake in a particular midwest or eastern United States region may be a long time.

Scott: Yes. It has always been a puzzle to me as to what ought to be done in such areas. The next earthquake may not come for many centuries. On the other hand, it may come sooner, and it may be a large, full-fledged earthquake. What do you do? Do you ignore the potential threat? Do you design conservatively, as if there might be a damaging earthquake soon? Do you do some seismic design, but with less conservative requirements? It seems like an awkward matter.

Crandall: You are right, those long-range occurrences are up for grabs, really. The doggone earthquake could happen tomorrow. That is where the uncertainty comes in.

That is why I think the buildings need to be designed as if the earthquake was going to happen the day after construction is finished. You might, however, be willing to take a little more risk in an environment where the seismic hazard has a long return period. San Diego, for example, can be expected to experience major earthquakes less often than Los Angeles. And the general soil characteristics in San Diego are better. Seismologists may point out that there is the Rose Canyon fault, and it is more active than we thought, and it could generate about a magnitude 6.5. But even so, you can say the risk of strong shaking is significantly less in San Diego than it is in Los Angeles. And then you can say the risk is a little less in another locale, and so on.

In the midwest, the New Madrid earthquakes happened in 1811 and 1812, and shook the bejesus out of the whole area. What should they do there? Such large earthquakes are estimated by geologists and seismologists as happening on average every couple thousand years, which is the mean or average return period. In most of California, you can get severe shaking with a much shorter average return period, much more frequently. That's the same as saying there is a higher probability of getting that severe shaking over the next so many years in California than in the midwest. The midwest needs some provisions for design, however, and they are just now getting around to believing that they ought to be doing something. In the Memphis and St. Louis areas—as of now, at the beginning of the 1990s—for example, they have been very slow in picking up on this.

Base Isolation of Structures

Scott: Talk a little bit about base isolation.

Crandall: Base isolation is a rapidly emerging technology for protecting structures from the full impact of ground motion. It has been discussed in concept for many years, but only relatively recently has the technology been developed sufficiently to permit its use in actual structures.

My firm was fortunate enough to be one of those involved in the first major building in the United States to use base isolation. That was San Bernardino County Foothill Communities

Law and Justice Center, which is the brainchild of the late Robert Rigney.[10]

Rigney was the county chief administrative officer at the time, and formerly a chairman of the Seismic Safety Commission. I was on the Commission at the same time Bob Rigney was, but was not on it under his chairmanship. He had completed his term as chair by the time I came on the Commission. It was certainly an honor and an achievement for Bob Rigney to have that structure built to a base isolation design.

Scott: I guess he was almost single-handedly responsible for getting it done that way.

Crandall: Yes. The funding was the problem. Being the first structure of its type, the county building department was very uncertain of the design. The savings that might otherwise have been realized were not possible because the building was required to meet the standard building code, to be designed for the usual high force levels instead of the lower level the isolation system would provide. So instead of there being an approximate equivalence of cost—if isolation systems had been treated the way they are now, with a realistic estimate of how much lower the forces will be—or perhaps even a net savings, that double requirement [of meeting both code standards and providing isolators] actually increased the cost significantly. Well, Bob Rigney, bless his heart, he is no longer with us, managed to convince the authorities and others in San Bernardino County that this

---

10.   This base-isolated building was designated as one of the Top Seismic Projects of the Twentieth Century by the Applied Technology Council and American Society of Civil Engineers in 2006.

was a good thing to do anyway. So the building was designed in that manner.

For a frame of reference, my firm provided the foundation design information and the ground motion design criteria based on the potential damage from an earthquake on the San Andreas fault, which is very near, about ten miles away, I think.

The basement area under and around the building is fifteen inches on each side larger than the building itself. In other words, the assumption was that the ground could move fifteen inches, and if base isolation worked, the building would lag behind that motion and it would reduce the forces in it to a relatively small amount, but you had to have the clearance. The foundation of the building itself is expected to move with the ground, to move fifteen inches each way in a major quake, but the movement would be absorbed in the isolators, so the building above those isolators should move very little. That's the principle of base isolation. It isn't total isolation, but in essence, the ground moves, and the building stays put.

Actually, of course, the building does move some, but much less, which means that the seismic forces in the building are very much reduced. If a building is designed in accordance with that principle, it does not need nearly as much strength or ductility as it would without a base isolation system. That can save money, and the performance can go up.

The isolation concept has been known for a long time, but they did not have the type of devices or the confidence in them that we now have, which permits us to do this. Earlier designers thought about putting buildings on

ball bearings. That would, of course, accomplish something similar, but the problem was whether bearings of that type would function and accept such large movements. Also, how do you stop the building if it starts going?

Anyway, this particular structure was the first major building of its type in the United States. Base isolation had already been used on bridges and a few buildings in New Zealand, where it originated. Now we have additional base isolated buildings in the USA. For example, the new University of Southern California Hospital is a base isolation structure. That shows you how far we have come, to do that with a hospital, which is subject to the most intensive seismic code requirements because it is intended to remain operating after an earthquake. So that is being done, and the hospital is expected to perform properly. Crandall and Associates was the geotechnical firm for that hospital design.

Those advances in the construction phase of engineering are very important for earthquake resistance. By that I mean we have designers who have long dreamed of things of this kind, but putting them into practical application requires the construction elements to be defined and developed. The University of California at Berkeley played an important part in testing these isolators, to show how they can function and provide great confidence.

Scott: Yes. Berkeley civil engineering professor James Kelly was one of those involved.

Crandall: Yes. That's right. Kelly was the key researcher. On the Law and Justice Center project, Alex Tarics was the key structural engineer.

Scott: In trying to follow the discussion of base isolation over the past ten or fifteen years, I tended to listen a good deal to Henry Degenkolb. He tended to be a bit skeptical, or at least cautious, about base isolation. Are you aware of that?

Crandall: Yes, I am.

Scott: Do you want to comment a little on Henry Degenkolb's concerns, as you understood them? Was it partly just caution on his part, because base isolation was such a new thing? Maybe he wanted to see it tried out in an earthquake or two before finally making up his mind?

Crandall: Henry was the kind of guy who could cut through all of the malarkey and monkey business and get right to the guts of something. And he was an advocate of redundant design—systems that had enough paths for the forces to flow along, so that if one of them turned out to be weaker than you expected, there was something else to help out. And of course, base isolation was a new concept. I had felt the same way about it.

You are reluctant to promote something of that sort until you have a degree of confidence, which we now have. I think Henry would have accepted the type of behavior that was exhibited in the test that Kelly ran as being indicative that base isolation would do what people hoped it would, and what the promoters of it were convinced that it would. You know how promoters are—everything is rosy to them until something goes wrong.

I think Henry just felt that good engineering was based on good solid construction

techniques. He was right about that, there is no question in my mind. But once you have shown what this system is capable of doing, then it goes into the designer's tool kit—one of those things that you can use. You still design a solid, earthquake-resistant structure, but it will be exposed to lower forces.

Scott: Then my recollection dates back to an earlier period when Henry was still wanting to be shown more proof that the new idea of isolation would work, and when we probably had not yet quite advanced far enough in demonstrating its performance to his satisfaction. Is that correct?

Crandall: I think that's right. Had he lived longer, he would have been in the era when isolation technology became something the engineer could routinely trust, if it's done right.

Site-Specific Ground Motion Studies

Crandall: The Uniform Building Code says that most of California is in the highest seismic zone, Zone 4. Well, we don't have to be earth scientists to know that Zone 4, which is most of California, has conditions that vary from the granite of the Sierra to the fill and Bay mud conditions of the Marina area in San Francisco. It is obvious that buildings of similar character on sites of those two extreme conditions are going to behave a heck of a lot differently.

If the example of Zone 4 and the other UBC zones are considered as zoning, then microzonation is mapping and studying expected ground motion at the scale of districts and individual building sites. That is what the soil engineers do when they make an investigation. They are doing microzonation for the specific site they are investigating. A soil or geotechnical report should include the geologic characteristics of the site and its subsurface conditions, such as whether it is on man-made fill or on Bay mud. So in essence, when you as a client retain a soil engineer, that consultant is microzoning that particular site for you. Or at least that consultant should, for a building of any consequence.

But beyond that, microzonation maps are being prepared for whole regions, and are becoming more and more refined, identifying such areas as—using the Bay Area as an example—the younger Bay muds around the Bay where man-made fills are, or where liquefaction is a possibility.

Scott: Who makes those maps? The U.S. Geological Survey, USGS?

Crandall: Yes, they do. The maps are also done by the California Division of Mines and Geology, or a city or a county. The seismic safety element required in local government plans in California has encouraged this sort of thing.

Computers: The Biggest Influence on Earthquake Engineering

Scott: Here in 1991, a number of "themes for discussion" are listed in the advance program of the Second International Conference on Recent Advances in Geotechnical Earthquake Engineering and Soil Dynamics, to be held in March 1991, ten years after the First International Conference in 1981. I thought it might be helpful for you to comment on these themes in discussing the recent and current

state of the art in geotechnical engineering, and in comparing and contrasting it with the state of the art in earlier periods.

Crandall: This listing in the advance program is a very comprehensive review of the field from the perspective of someone who started out in the subject in 1940-1941. There are things here that we had not considered or thought of in those early days. One example is the extensive discussion of computer programs at these conferences now.

I think that much of what is being done now is the result of computer capability. When you are trying to do things by slide rule, which we did, of course, in those early days, you find a limit to the extent of mathematical analysis that you can do. What we did not know in those days we approximated by assuming boundary conditions. In other words, we took a low value and an upper value for the shear strength, and computed what would happen.

Suppose we were considering settlement of a structure, or the motion a foundation would undergo if subjected to the load of a generator or some other piece of rotating equipment, or even of an earthquake. We had, back then, to estimate the lowest value and the highest value of the shear characteristics that might be anticipated for that particular installation. Then we would try to determine what kinds of deflections would occur under each of those values, and somewhere in between, you hoped, was the correct answer. If the spread was too great, then you either had to throw out your whole process and fly by the seat of your pants entirely and use your judgment only, or try to refine the bounds and improve on the first estimate. Methods like the response spectrum were made feasible for the consulting engineer by the modern computer.

The limitations on analysis before then were what you could do on a hand calculator or a slide rule. As a result, you could not make many iterations of all the calculations. You took just one or two shots at it, and said, "All right, we think it is going to be one-half an inch." Now, of course, the process is greatly refined, although maybe in some cases, the answers aren't any better. I sometimes have doubts. The mere fact that you have a computer that will spit out all of these calculations to ten decimal places does not necessarily mean you get the right answer. One of the toughest things I find in hiring young engineers just out of college is teaching them not to accept the results they get in computer output on blind faith. The input is, of course, what controls the validity of the results. It is too easy to look at a lot of significant figures and say, "Oh boy, this must be right because we are measuring it to a thousandth of an inch." But maybe you are really only measuring it to one inch with accuracy. Before the computer, we could not do as much analysis as now, but we benefited from the fact that we knew how important judgment and experience were.

Scott: When did that transformation really begin? I guess it took a while to really influence the profession.

Crandall: I would say it was in the mid-1960s and thereabouts that the computer began to be a useful tool in engineering offices.

Scott: Of course, then there were successive generations of computers and, they got cheaper and cheaper, and their abilities greater and greater.

Crandall: Yes, they could do more things in a shorter period of time. The explosion of computer knowledge was just tremendous. The development of software, for example, as people began to work on these problems. Today, my goodness, you can get a software program an average office can utilize with a computer, and put in the basic knowledge, and even more than that—make alternative studies of the effect of different configurations, of different connections and bracing systems. That way, one can select the most advantageous system and come up with some economic factors to indicate what the additional cost would be of improving the design.

In the early days just doing the fundamental soil engineering, computing settlement of a building foundation, for example, was a laborious hand calculation job. As a result, we did not do too much of it. You made typical foundations for a building, made a calculation, and then you extrapolated from that to other sizes of foundation. It was well known at that time that the size of the foundation and the load or the pressure on it, say of a spread footing, was very critical in the settlement determination. Once you knew the soil characteristics, then, of course, you had to make some evaluations of what settlement was going to occur, and particularly the differential settlement between, say, a major foundation element in a building and the nearby lighter load-carrying foundation element of the building. Those things were based on a lot of seat-of-the-pants type of judgments. Not that they were wrong, but it required more experience to do that. Nowadays, it is somewhat different.

Scott: Did the computer affect both professions, both structural engineers and soil engineers? And these effects intertwined?

Crandall: Yes. I think the use of computers for dynamic analysis started with the structural engineers. They were able to take the lateral forces and evaluate them in an economic manner. Once that happened, they started asking questions about the effect of the ground motion on the structure.

The structural engineer says, "I'm analyzing the behavior of the structure itself to a degree that I had never been able to do, and now I'd better find out something about the ground motion, because that affects it." The geotechnical types, the soil engineers, were in a sense forced into determining the behavior of the ground, so that the structural engineers would have that information to utilize in their design.

Advent of Large Shaking Tables

Crandall: Another development that started to come in around the time when computers became useful to consulting engineers, in the 1960s and 1970s, was the large shaking table testing capability. There are these large-scale and full-scale capabilities for testing mostly structures, but also for testing soil. We can now do soil tests for very large specimens on shaking tables. That gives you much better, firmer results than before.

Scott: You mean instead of testing a small sample of soil, you test a huge amount? Give me an idea of the size, the magnitude of the specimens you have in mind when you say

"large-scale." Do you mean a six-foot cube, for example?

Crandall: Yes, that is in the realm. The typical soil specimen is maybe two or three inches in diameter, and maybe six to eight inches in length. We could prepare larger specimens— "remolded" we sometimes called them, or "reconstituted"—in say a one-cubic-foot box, and do some tests. But we could not give it the kind of dynamic testing that we can now do with a shaking table.

Scott: When you say "shaking table" are you referring to one like the one that the University of California at Berkeley started to operate at its nearby Richmond Field Station in 1972?

Crandall: That is available and has been used for some soil tests. But usually smaller tables would be used in soil engineering. I'm thinking of tables maybe four feet square to six feet square.

Scott: Are there many of those around?

Crandall: There must be a dozen or so in California at the various universities. Maybe even some private soil laboratories have equivalent types of facilities. Mine doesn't. We haven't gotten into it to that degree, because of the cost. We would just take our samples in to UCLA and contract with them to make certain special tests. It makes sense. They can stay up with the state-of-the-art, whereas a private firm could not afford to change the facility as improvements come along.

Even the shaking table is now a function of the computer, which is set up to have two-dimensional motion, two horizontal axes, as controlled by the computer program. Some of them even include the third component, the vertical shaking. These motions of the table are all controlled by computer programming now. You can duplicate an earthquake—any specific recorded earthquake motion—in the shaking program and play it back on the table.

That has been a really big advantage in helping understand the dynamic properties of soils. Since all buildings are supported on foundations, understanding what happens to the soils or the underlying material gives the input that affects the building itself. A lot of people have forgotten that the effect of an earthquake actually comes from the ground into the building.

Scott: Do engineers still need real earthquakes to provide them with "test results"?

Crandall: Yes. An earthquake is a real, full-scale test of a building, and unfortunately, we still need that type of test to verify and advance our field. Shaking tables test models. Even the biggest shaking tables cannot take a big, full-sized building. There are things that go on in a real building that you cannot reproduce in a model. For example, the interior partitions, the walls, the floors, the stairways, and things that make it very complex in trying to make a calculation. Then there's the behavior of the soil that we have discussed. The actual measurement through strong motion instrumentation lets you have a good feel for the effect of these nonstructural elements.

Measuring Shear Wave Velocity

Crandall: One of the things I should talk about is early studies of shear wave velocity. It is one of the simplest parameters in soil dynamics, but an important one. Enough stud-

ies have been made in earthquake engineering to know that the behavior of a particular site is a function of the shear wave velocity of that material, the density of the material, the depth of bedrock, the variation of the soil layers beneath the site, and such things as the distance to the focus of an earthquake. Also, of course, the magnitude of the energy generated in the waves arriving at the site is important.

Initially, shear wave velocity was estimated from laboratory test data, but we soon realized that was a very crude procedure, and so we started measuring site vibration characteristics with geophones. Initially, this was done on the surface, measuring the time required for an impact at a source to reach a series of instruments that would detect the motion.

Scott: Using sensitive listening instruments?

Crandall: Yes. They would pick up the vibration, and the time could be measured to a fraction of a second.

Scott: How did you produce the vibration?

Crandall: With a sledgehammer. You hit a plank that was weighted down by the wheels of a truck, so the plank was in intimate contact with the ground and the vibration from the sledgehammer hitting went on into the ground. The time that the sledge hit the plank was time zero. The various waves traveled through the ground to the string of geophones that were set out at varying distances from the impact source.

The geophones recorded with a precision that would permit us to determine to a thousandth of a second—a millisecond—the arrival times at the various geophones. From that, we determined the velocity, because we knew the only other variable, the distance. Both the P wave and the S wave velocities were measured. This was later improved by making a downhole measurement, in which the geophone was lowered into a boring at depth, and the sledge-hammer was struck on the surface.

Scott: How deep was the geophone placed?

Crandall: The geophone was lowered to varying depths. Some of the borings would be 200 feet deep. You would start by putting the receiver down at the bottom of the hole, hit the plank at the top, and measure the time for that geophone to receive that wave. Then you would raise that geophone to say 150 feet depth, and do it again. You'd keep doing that successively as you withdrew the geophone from the hole, getting a time-of-travel determination from which you could determine both the compression wave (or P wave) and the shear wave (or S wave), as they were called. Knowing that, you had information that could be used to determine what the effects of a distant earthquake would be on that site, in terms of that velocity measurement.

It was back about 1965 that Martin Duke of UCLA worked on this geophysical procedure. He got it from the Japanese, and brought it to the United States. He had his students working on it. Crandall and Associates provided the borings and the other material, and Martin had the expertise and crew to do the technical work.

Scott: Were these experiments done on actual commercial jobs?

Crandall: Yes. The first one I remember was the Music Center for the City of Los Angeles. The Music Center was located on what we loosely refer to as "rock" in Los Angeles—the siltstone formation or shale, as it's generally called, of the Bunker Hill area of downtown Los Angeles. Those experiments were the first observations that I recall being made using that procedure with the geophones and pounding on the plank with a sledgehammer.

We collected the data, knew the shear wave velocity, and made some feeble attempts to predict how much the ground motion might be. The first thing we did was determine the natural period of the ground. That is what we were originally looking for. The period of the ground could then be related to the period of the structure.

The structural engineers were now capable of determining the period of a structure rather precisely, rather than using the arbitrary formulas employed before, when they just estimated it from the number of stories of the building. We, geotechnical engineers, were then able to provide what we felt was good information of the natural frequency of the soil. This was one of the early factors required in the dynamic analysis of a structure.

Response Spectrum

Crandall: Computer developments permitted geotechnical consultants to calculate the earthquake spectrum, which combines information on ground shaking severity as compared to its frequency. The response spectrum plots the maximum response of a variety of different structures, each of which has its own period of vibration.

So the process used then was to estimate how big the earthquake might be, figure its distance from the site, and use that information together with the known characteristics of the shear wave velocity of the site, the depth to bedrock beneath the site, and the type of material, and prepare a very simple response spectrum for that particular site. With that information, the structural engineer, knowing the period of his or her building, would have an estimate of what the predominant period of the ground motion would be, and the acceleration, the velocity, and the displacement of the ground for a given earthquake.

The structural engineers could then use this in their computer programs. Again, I give much credit to the computer. The structural engineer was able to use the response spectrum in studies of the effect of the ground motion on a particular building. Since those days, the process has been refined considerably, but essentially it's the same information that is determined and prepared.

Defining the Design Earthquake

Crandall: In addition to the properties of the site, the other key factor was the choice of the earthquake. There are various possibilities, of course. Actually, the possibilities are virtually innumerable. Nobody knows exactly what earthquake will occur on a given fault.

Scott: Is that when such concepts as "maximum credible" and "maximum probable" came in?

Crandall: That's where all of that terminology began to enter the picture. It's well enough to be able to say that the ground motion will

be so much, and of such a character if this earthquake occurs. But then somebody says, "What's really likely to occur? We want to design this building for what may actually occur."

The earthquake engineering fraternity—which consists of more than just structural engineers and includes the seismologists, geologists, engineering geologists, geotechnical engineers—needed to come up with a series of determinants as to what energy might be released in earthquakes on a given fault.

This is where "maximum probable," "maximum credible," and even the "ultimate possible," and "maximum possible" terminology came up. The first effort came up with the "maximum possible" earthquake activity—it was almost an infinite-magnitude sort of thing. You had no limitation, really, so that was rather quickly dropped.

Scott: You wouldn't use that concept, unless possibly with a nuclear reactor.

Crandall: Right. They, of course, were being designed for a higher degree of uncertainty than the structural engineers of buildings were designing for. But for building design, it was finally decided that the "maximum probable" and the "maximum credible" were two levels of potential activity that should be considered. As a result, the geotechnical engineer typically ignores the "maximum possible," and develops ground motion characteristics for various faults for the "maximum probable earthquake," which gives a lower number than the "maximum credible."

In the Los Angeles area, for example, we always consider the San Andreas activity on a given site within, say typically, forty miles from the fault. So the behavior at that distance is much different than if you're close to that fault. The Newport-Inglewood fault, which is directly below a large part of Los Angeles, is the other fault considered. Then depending on the specific location under study, the Santa Monica-Hollywood fault or other faults in the vicinity are also considered, and the ground motion due to those faults predicted.[11]

Scott: They consider these two big ones, San Andreas and Newport-Inglewood, and then the others would be of more localized concern?

Crandall: That's right. Like in Pasadena, the Raymond fault is usually considered. In Culver City, the Overland fault is one. So you evaluate the geologic characteristics of a given area, and decide which faults might have an effect on your site. You then attempt to determine the ground motion under the possible event that you expect—either the maximum probable or the maximum credible. Those numbers have been pretty much developed by seismologists now as a function of the length of fault.

George Housner had much to do with that determination, in which the magnitude or the energy released on a given fault is related to the probable length of rupture. The longer the

---

11. Earthquakes that occurred later than the date of this interview with Crandall, such as the 1994 Northridge, California earthquake, were to also bring attention to the hazards of blind thrust faults in the Los Angeles region. Unlike faults such as the San Andreas or Newport-Inglewood, which provide geologists with evidence at the ground surface of their past displacement, "blind" or buried faults that do not extend to the surface are now known to be a significant hazard in the Los Angeles region.

rupture, of course, then the greater the total amount of energy.

Scott: You say George Housner was one key person in this?

Crandall: Yes. As I remember, George Housner came up with the first relationship of length of fault to the potential magnitude of an earthquake on that fault. Things are pretty much in that state of the art at the present time, although we are refining matters. Now we're considering the depth of the focus of an earthquake as important, with a shallow quake producing a certain kind of ground motion that is different than that of an earthquake at greater depth, and then it also matters whether it's a thrust fault or a strike-slip fault. Those are refinements that are being given consideration at the present time.

I feel that the state of the art now is such that we have a good degree of confidence in the data we're developing and are presenting to the design profession, namely the structural engineer and the civil engineer, for use in the determining what the effects of the ground motion will be on a particular structure. Again, I don't want to say this too many times, but this has become possible only because of the use of computers in the determination of these very complex factors.

Scott: I guess the development is still going on, although in some ways it may have more or less reached a plateau.

Crandall: Yes, I think it's flattening out. Now we have the ability to make these analyses. It's just a question of obtaining the input data, refining the data, and improving the determination of these various factors—plus, particularly, our understanding of the total amount of energy released.

Next Breakthrough: Total Energy, Number of Cycles

Crandall: One thing not yet routinely and explicitly considered as a part of seismic design is the duration of the shaking. That has always struck me as very, very important and fundamental. The initial efforts concentrated on acceleration, and we've talked about peak acceleration. The response spectrum gives you one peak value for a structure of a given period. But it's a well known fact that one or two cycles of, say, 50 percent g acceleration, are not going to cause serious problems in a well-constructed building. But if that 50 percent g keeps up for five, ten, or twenty cycles, then it has a tremendous effect on the building. It is a question of the total energy, rather than a single shock. It's recognized that duration is important, but to my knowledge there has not yet been any specific way of including the duration effect, other than as a judgment in the analysis and design of structures.

I have long maintained that what really counts is the total energy imparted to a building, although that is not easy to figure. It is not just the single maximum acceleration effect, but also involves the duration of the accelerations. This is recognized by almost everybody, but is not an easy thing to deal with. To my knowledge, the engineers have yet to succeed in thinking of an earthquake's effect in terms of total energy release and its impact on a building.

We measure earthquakes that way, of course—the Richter scale for example is really a measure of the energy release of earthquakes. The Richter magnitude is on a logarithmic scale of ten. You always read that in the newspapers, but the factor of ten applies to the amplitude of motion as measured on a seismograph. There is a factor of ten between the amplitude of a magnitude 5 and a magnitude 6 earthquake, as it reads out on a particular kind of seismograph, with a correction for the distance of the seismograph from the earthquake epicenter.

In terms of total energy released, however, the ratio between a magnitude 5 and a 6 is more like thirty-two times. So a magnitude 6 involves thirty-two times the energy that a magnitude 5 has, and a magnitude 7 involves roughly 1,000 times the energy of a magnitude 5 (32^2 equals a little over a thousand). The total energy released goes up very fast as you go up the magnitude scale. And that is related to the effect on a building, as well as the area affected by shaking.

My feeling is that the next breakthrough will be in the manner of considering what I will call total energy. That is, the number of cycles of a given level of shaking, and how they affect the building. We know very well that after a few cycles, the period of a building changes. And, of course, the strength of the connections and so on are affected. I think what that does to the building is very important, but at the moment, at least to my knowledge, it cannot be adequately and thoroughly considered in the design procedure.

Scott: That seems like a very important point. Isn't that one of the big unknowns about predicting future earthquakes—anticipating how long they will last?

Crandall: We know empirically, based on past earthquakes, that a low-magnitude earthquake is going to shake for an estimated length of time. That's about as good as we have done up to the present time. But that has not yet been worked into the design analysis.

Structural engineers have used different factors, based on the response spectra and the ground motion interpretation. I think the next step in engineering analysis of buildings will be the methods of including the energy consideration, the number of cycles of shaking of a given level.

Scott: What will it take? More actual observations in earthquakes made with strong motion instruments and other kinds of instruments? Shaking table experiments? Theoretical computations?

Crandall: Those are all important, and they will give a better understanding of how this thing happens. I think, however, that (hopefully) computer programming in the near future will be capable of analyzing a building with an assumed number of cycles of shaking.

Scott: In other words, the designers will successively plug in a bunch of different assumed earthquakes, and see what happens?

Crandall: Yes, that's the basic idea.

Lifeline Earthquake Engineering and TCLEE

Crandall: We've talked about buildings and seismic design, but dams, bridges, power plants, storage tanks, and other civil engineering works are also important. At this point I might discuss another element of earthquake engineering that I had a hand in, and that was the Technical Council on Lifeline Earthquake Engineering, TCLEE, which is a council of the American Society of Civil Engineers.

Scott: I remember Martin Duke using that acronym at EERI conferences. At first I didn't know what it was all about.

Crandall: Martin is very central to the story. The whole thing started on a plane ride from Sacramento to Los Angeles. We were returning from a Strong Motion Instrumentation Advisory Committee meeting. Martin Duke and I were seatmates on the plane going back to Los Angeles. He is now departed, having passed away in 1988, and we miss him greatly. I was on the board of direction of ASCE at the time. Martin was talking about all the seismic effort put on buildings, and yet there were many other important structures that should be thought about, namely dams, bridges, highways, communications systems, and so on. These were all matters of the sort that Martin lumped under the term "lifelines." I don't know if he originated the name or got it from somewhere else, but I first heard of it through Martin. I'll give him the credit for being the first to at least apply that name in our earthquake engineering field.

Scott: It is almost universally used now.

Crandall: Yes. I don't know what the exact origin is, but certainly Martin made it popular. Anyway, we got to talking about the matter on the plane trip, and Martin said he felt there was a big need for information on these civil engineering works that affect the safety and services of the public. We talked for a while, and then the thought came to me, as Martin expounded on his thinking, that the civil engineers who design and operate these facilities should be concerned with this.

It occurred to me that the American Society of Civil Engineers would be a logical place to develop an organization with this subject as its principal topic for consideration. Martin thought that was an excellent idea. He was also active in ASCE. Martin and I talked about it and came to the joint conclusion that it would be desirable to set up a committee to explore the possibility of having a division of ASCE to consider lifeline problems in earthquake engineering. The ideas jelled, and I presented the concept to a meeting of the board of direction heading up ASCE. The procedure set up for starting new activities was, first to have a committee to study the matter, see what its scope was, and estimate the interest of the society members. I obtained the go-ahead from the national board for that.

So we set up a study group. Martin Duke, if my memory is correct, was asked to serve as chairman, and I was the board of direction member appointed to be the liaison between the new committee and the national board of direction of ASCE.

We hand-selected a number of people we felt had an interest. These were people from the public utilities—for example, the Edison Company of Southern California and the Los Angeles Department of Water and Power. Well, the committee met several times and decided that there certainly was a need for this, especially among civil engineers. A survey was made of the society members to see what their

interest was. The response was very gratifying. The next step, then, after the committee recommended that ASCE take action in this field, was to establish a technical council.

It was called the Technical Council on Lifeline Earthquake Engineering, TCLEE. The format of this group was set, and I was privileged to be the one to introduce the motion and resolution to the ASCE board to go ahead and establish the council.

Scott: Roughly when was the technical council established?

Crandall: It was in 1974, after the San Fernando earthquake. Martin was EERI President from 1970 to 1973 and was very involved in all the studies launched after that earthquake, including his interest in lifelines. At first, TCLEE was under the umbrella of one of the ASCE technical divisions. Later, because of the number of members interested, it became a full formal technical council. It has been operating at that level ever since. It has its own publication, and every year or two holds its own meetings in conjunction with national ASCE conventions. TCLEE sponsors part of the convention program. Very good work has been done and disseminated by that organization.

Martin Duke was the parent of TCLEE—it was his basic idea. I participated in the sense that I came up with the approach of doing it through ASCE, and due to my being on the board as a national director, was able to introduce it and bring this to an early establishment.

Scott: Looking back, has TCLEE accomplished most of, or even more than, what you two had in mind when you first discussed it?

Crandall: I think it has grown well beyond anything we thought possible. In just one field, pipelines, and how to design a pipeline crossing a fault, for example, there have been numerous studies, and NSF grants to research institutions for evaluating the various factors. Shaking is normally of no consequence to pipelines, because the pipeline is moving with the ground, and the distance over which the shaking takes place is long enough to allow a pipe to bend and accommodate any differential soil motion. The wavelength of most of the earthquake waves is hundreds of feet. So from one point where the ground motion is going one way to the next point where it is going the opposite way, there is usually a sufficient distance that it has no effect on a buried structure. The difference in the way the soil is moving is spread out over a long distance of a pipeline, for example.

But, of course, if there is a sharp break in the ground, that is another story. The Bay Area Rapid Transit (BART) tube, for example, has a comfort zone built into it so that a displacement of several feet where the fault crosses the line can be absorbed in the tunnel. In other words, a kind of a joint was built at that point.

We have done the same thing on a major sewer project in Los Angeles that had to cross the Newport-Inglewood fault, the North Outfall Replacement Sewer (NORS) project. We consulted with Clarence Allen about the amount of displacement that one might expect in the event of a rupture of the Newport-Inglewood fault at the location of our sewer line. If I remember correctly, together Clarence Allen and we decided that three feet of possible lateral

movement, and of course some upward movement, could occur there.

So we built a soft zone, like a pipe within a pipe, across that fault. The outer pipe is expendable. The inner one is supported free and clear with a sufficient distance on each side so that if the ground moved the sewer line, although it would still have to bend, would bend to accommodate a few feet of bending over a length of 100 feet, say. The outer shell would break, because it was a tunnel through the soil and would be sheared over a short length where it crossed the fault. The inner pipeline could absorb the distortion over a long distance.

You cannot do much for a trenched pipeline in city streets where there is going to be liquefaction, with very abrupt changes where the pipe goes from solid ground into soft ground. The joints of those old pipelines are very brittle, so it would not take very much to damage a buried pipe under such conditions of discontinuity. If there is no soil discontinuity, then they are almost invulnerable to such problems, in my opinion, because they move with the surrounding soil.

Highways and bridges are important lifelines also. What happened to the Cypress Viaduct in Oakland in the Loma Prieta earthquake was due to amplified ground motion, primarily because the structure was of an early design that it is now recognized as incapable of resisting this kind of shaking. But that does not mean that all freeway designs, especially the newer ones and those built from this time on, are going to have the same experience. A simple bridge structure is the simplest thing there is to design. You know the loads beautifully, the geometry is elemental, supported on a couple of columns at each end. As far as design goes, it is much less complex than a multistory building. We are certainly able to design and build that kind of bridge structure, whether one deck or two, with any degree of safety that is wanted.

Scott: So banning two-level structures is learning the wrong lesson? The real lesson is to do a better job of designing two-level structures, not to avoid them entirely.

Crandall: Yes.

Parkfield Experiment

Crandall: One example is the experimental pipeline installed in the Parkfield, California area along the San Andreas fault, halfway between Los Angeles and San Francisco. The pipeline crosses the fault and is instrumented, awaiting the predicted earthquake in the Parkfield area.

I should mention that the principal geotechnical engineering firms in California —and I was chairman of the committee—have gotten together and made instrumental readings in the Parkfield area at a place called Turkey Flat. Exploration borings were drilled along this site. Various firms participated in that under the auspices of SMIP, the Strong Motion Instrumentation Program of the state of California, and the California Division of Mines and Geology.

We got together, and based on the data provided, each firm predicted what the ground motion would be at various points adjacent to the fault. SMIP has established instruments at those points. The idea behind this was to have a full-scale test of how accurate we are in our predictions.

Scott: If and when the predicted Parkfield earthquake does happen, you're ready for it with those various guesstimates and predictions.

Crandall: That's the idea. These were predicted in advance and submitted, so nobody can say, "I didn't really mean that." When that earthquake happens, we'll see what procedures are best, what worked the best, and which firms are doing a job that gives the best answers. This is a very important step, because you do all these things with the computer—you do calculations based on measurements and you extrapolate. Here is a case where we will get what we think will be about a magnitude 5.5 or 6, something like that, in the near future. The instrumentation is in place. The predictions have been made.[12]

---

12. On September 28, 2004, a magnitude 6 earthquake occurred on the Parkfield segment of the San Andreas fault. See Real, C.R., Shakal, A. F., and Tucker, B.E., "Turkey Flat, U.S.A. Site Effects Test Area: Anatomy of a Blind Ground-Motion Prediction Test," Third International Symposium on the Effects of Surface Geology on Seismic Motion, Grenoble, France, 2006. In this context, "blind" refers to the fact that the people making predictions of the strong ground motion could not see any data from it, since they made their predictions in advance of the occurrence of the earthquake.

Chapter 9

Contributions of an Older Generation

When something goes wrong, the owner gets the lawyers to put the noose around anyone they can find, and with the developer long gone, often the soil engineer is who they find. We're in a terrible business from that standpoint.

Clarence Derrick: A Unique Engineer

Crandall: I'd like to say a few words about Clarence Derrick, who was a very talented structural engineer of the old school here in Los Angeles. He was kind of a mentor of mine, and took an interest in my career. He was very helpful in giving advice, for which he was quite famous. Clarence would give advice on almost anything, whether you wanted it or not.

I recall with great pleasure having the privilege of knowing Clarence Derrick. Clarence practiced in Los Angeles and was a graduate of Notre Dame. He was a unique engineer in that he had a very broad background, including literature and the arts. He was not an artist himself, but he was familiar with them. It must have been due to the education he received at Notre Dame. He must have graduated from Notre Dame about the time of World War I.

He was extremely literate, and he felt there was no school like Notre Dame. Clarence was an Irishman. He was very cultured and would quote from the Bible and from literature. He was a very articulate, very impressive guy. He could also get down on the floor and shoot craps with you. Clarence could charm you very readily, but he was also a man of substance, particularly in engineering.

When I started my own business, Clarence used to say, "Do the job, don't worry about the money, if you just do things right, the money will come. But if you just set your goal for making money, you have the wrong sights and you very probably will be a failure."

Scott: You think that was very good advice?

Crandall: I know it was. I already felt that way anyway, but he reinforced my thinking. In my practice we have always ignored the contract amount and done the job the way we thought it needed to be done, and the way the client expected us to. If we made some money on it, that was fine, and if we lost, well that wasn't fine, but we accepted it.

We never ever shortchanged a job because we had made a poor cost estimate originally. I can't think of any case where we did not deliver what we thought the job needed. Now sometimes you estimate a job to begin with, and many clients want a flat fee. But you get out there and start drilling, and you hit something unexpected. What do you do? Oftentimes the client would be understand-

ing, and were willing to accept an increase in fee. But many times they kind of felt maybe it was our own fault, that we had not researched the job enough, or for whatever cause. Anyway, Clarence was a great philosopher, and that was one of the things he impressed upon me.

Clarence had knowledge about almost everything, and was a tremendous engineer who cut through a lot of the mystery of early seismic design. He wrote a couple of books that were not widely circulated, but were used in a course he taught at the University of Southern California (USC), which must have been in the late 1940s and early 1950s. He used a very simplified procedure for describing earthquake motion in the ground and as transferred to a building.

I also want to discuss Clarence's research interest in earthquake effects. At that time they designed pretty much by the code, typically with an equivalent static lateral force based on 10 percent g, or 20 percent g if you felt you had a really critical building. There wasn't much understanding of what was actually happening. Clarence, in his own basement laboratory at home, did some outstanding measurements of building behavior using models. He developed a small, one-directional shaking table. One of his models was a four-story frame made of aluminum, scaled at about ten inches per floor, about forty inches high in all, and perhaps eight inches in width. Another model was just a two-dimensional frame. He had two steel strips, maybe three feet high, and an inch wide and a sixteenth of an inch thick. Then he had floors connected across at appropriate levels, so the model was divided up into maybe six levels. The whole thing was something like ten inches in width, and some thirty inches high. With six levels, that would make it five inches between each level.

The frame was mounted on a platform that he could shake back and forth with a synchronous motion. The frame would sway as he made these motions. Engineers usually thought the building had just one mode shape—it bent over, and it bent back, it bent over, and it bent back. Clarence said, "No, that isn't right. There are other wiggles that come into it." That depended on the degree of shaking and the character of what he had built. He had it so he could clip additional weight to each floor, and make changes that way, to represent a heavier structure, compared to a lighter one.

He would subject the model to a shaking movement at the base. This was before we had high-speed cameras, or had ready access to them. He wanted to see how the model deformed at various intervals, in split seconds of timing. He set up his still camera with a switch. The switch was activated by a weight that dropped in a glass tube. You know from Newton's laws how long it takes to drop say two inches, five inches, ten inches, or twenty inches. Derrick came up with a timing mechanism that was incredible. The guy was a gun enthusiast—of the muzzle-loader type. So he took a lead bullet, and mounted a yardstick vertically alongside the shaking table. The bullet had a steel jacket on it, and a magnet held it up say three feet above the table. The bullet was in a glass cylinder, and would fall when he released the magnet.

At a certain distance down the yardstick scale, say maybe at the ten-inch point, he would have a wire that he could insert into the cylinder, so

when the bullet fell and hit the wire, the flash camera went off and took a picture at that precise moment. He did a series of pictures like that. He set the bullet, dropped it say five inches, took a picture, then dropped it ten inches, took another picture, and continued like that with a very accurate timing mechanism that was so simple it was incredible.

Scott: It would take one picture each time he dropped the bullet?

Crandall: Yes. Then he would advance the film, get the shaking going, and do it again. He came up with a series of photographs that showed the deformation, the S-type wave. He impressed and amazed engineers like Steve Barnes, Murray Erick, who was alive at the time, and Oliver Bowen. Also Paul Jeffers, who was as skeptical as hell about everything anybody else did. Clarence used that shake table and the photographs to train other engineers, younger people. This was before we had computers, probably in 1948 or 1949.

Scott: Is the experimentation with the model described in his writing?

Crandall: Yes. He wrote several that I still have. Here are the titles: *Damage Potential of Earth Shocks* (1954); *Elements of Aseismic Design* (1955); *Aseismic Design by Distortion Analysis* (1956 and 1957); *Elements of Aseismic Design, Part II* (1959). The engineering students at USC were exposed to those writings. When you saw these pictures in sequence and the deformation of the frame and how it was moving, it really gave a graphic illustration of what happens to a simple frame building during a shaking period.

He developed what he called a distortion analysis system, based on that information and his simplified ground shaking. I took his course and read parts of his book, but it has been a long time. I don't want to go into this too far, but his ground motion was in a semi-circle, a half-circle. Then it went in a straight line for a short distance, until it turned back again. It was kind of like an ellipse.

I guess it had circular ends, and was like cutting a pie in half and separating the two semi-circular halves by maybe six or seven inches. That was the shape of his description of the path of a particle at the base of the building. It showed the movement that would be transmitted, and was rather simply solvable mathematically, which was the key to the whole thing, finding a simple basis for determining what is the effect of the ground motion.

Derrick deserves more credit than he has been given.

Scott: I hear you. You know, from John Blume, principally, and some others who worked with him, I have gotten background on some of the work done up at Stanford. John built a model of a building, the Alexander Building, and at one point he and a fellow student did a painstaking analysis, using Marchant calculators, of what happens in such a building under earthquake forces.

Then from George Housner I got background on some of the activities mainly centered around Caltech. But nobody that I recall mentioned this work by Derrick. I believe Bill Moore did mention Derrick as somebody of consequence in earthquake engineering. But you are really the first person who has talked much to me about the nature of his work.

Crandall: Derrick came up with formulas for the effects of the vibration, the acceleration, and the velocity, and came up with a displacement formula. He then talked with the head of structural engineering at the University of Southern California, David Wilson. That is another story—I think Dave Wilson had one of the biggest impacts on civil and structural engineering in southern California at that time, because of his teaching ability and the inspiration he was for his students at USC.

At any rate, Clarence Derrick got together with Dave Wilson, who was so entranced by what Clarence was doing that he asked him to teach the subject at USC. So as I noted earlier, Clarence taught at USC. For about three or four years he gave a course in aseismic design. He was a stickler for language, and he said it should be "aseismic," not "seismic."

Scott: Meaning non-seismic.

Crandall: Non-seismic, yes. He got very few supporters on the use of that term, but strictly speaking, Clarence was right, you know. It is a little like the word "anti-seismic" that Stephen Tobriner used when he gave a talk for one of our Seismic Safety Commission workshops.

Anyway, Clarence taught at USC. I took one of the classes, even though I was not a structural engineer, because it had so much information for a soils-type person. He did that for a few years until he got some trainee who carried on, and he gave up the teaching because he had other interests.

When I knew him, I do not recall that he ever designed a building himself as a structural engineer. He had done that in the past. He was in on some of the major structures in southern California, and I think he even worked on the Los Angeles City Hall, which Albert C. Martin was involved in.

Clarence was in demand as a reviewer and consultant for many buildings. I know the designs of the county buildings at the Civic Center in Los Angeles, the Hall of Administration and the County Courts building, were reviewed and critiqued by Clarence for the architects. Brandow and Johnston was the structural engineer for the buildings, but Clarence Derrick worked with them in improving the designs and bringing into actual practice the type of thing he had worked up in his studies. Roy Johnston knows about Derrick.

Incidentally, Clarence Derrick was one of the founders of the Structural Engineers Association of Southern California, one of the "dirty dozen" as they called them. They were the guys who got together and originated the structural engineers association. It began in southern California.[13] That all began with Clarence, Paul Jeffers, Steve Barnes, guys like that.

---

13. The Structural Engineers Association of Southern California (SEAOSC), was established in 1929, followed by the establishment of similar organizations in other parts of California: the Structural Engineers Association of Northern California (SEAONC), of Central California (SEAOCC), and of San Diego (SEAOSD). The charter members of SEAOSC were Rufus M. Beanfield, Oliver G. Bowen, Wendell Butts, Ralph A. DeLine, Clarence J. Derrick, Murray Erick, Mark Falk, Paul E. Jeffers, R.R. Martel, William Mellema, Clarence E. Noerenberg, and Blaine Noice. See "History," SEAOSC website, http://www.seaosc.org/about_history.cfm.

Fritz Matthiesen and Strong Motion Studies

Scott: You've mentioned Fritz Matthiesen to me a few times. Say a few words about him. As I recall, he was with USGS.

Crandall: Yes. USGS had what they called the earthquake group, and he was in it. Initially it was headquartered in the Bay Area. Before that, he was on the teaching staff at UCLA with Martin Duke. Then he left for the Coast and Geodetic Survey, and then went with the U. S. Geological Survey when the earthquake program shifted there. Fritz was in charge of all the USGS strong motion stations. At that time, USGS was practically the only one who had them set out in far-flung places in California, such as in the El Centro area in the Imperial Valley.

Fritz was in the forefront of gathering this type of strong motion knowledge, and was very outspoken and vehement about it. Fritz knew that strong motion data were essential to understanding and designing structures to resist earthquakes. He had some interesting characteristics. One was, he never wore a tie to any of the conferences that we were always going to in those days — meetings of the engineering groups and EERI and that sort of thing. I recall one year when they made a special surprise presentation to Fritz. They called him up to the head table and presented him with a tie, which he only wore that one evening.

Romeo R. Martel of Caltech

Scott: Take a look at these comments by Bill Moore in his EERI oral history. He said "Martel's work on moment distribution and analysis should be mentioned." Do you know anything about that?

Crandall: I only once had the pleasure of hearing Professor Martel speak at a meeting. Then he left the scene. He was a structural engineering professor at Caltech, and taught both Bill Moore and Trent Dames.

Scott: Do you recall when you heard Martel?

Crandall: Yes. I started with Dames and Moore in December 1941, and joined the Structural Engineers Association of Southern California. The company was very supportive of my doing that. They encouraged me to become active in structural engineering through SEAOSC, and in civil engineering by membership in ASCE. I was secretary of the Los Angeles section of ASCE in about 1945.

The Los Angeles, or southern California, section of the statewide Structural Engineers Association of California used to meet once a year at the California Institute of Technology, the next year at UCLA, and the next year at USC. On the years when we met at Caltech, Martel gave the address. I did not know him, but he was a distinguished-looking guy, very impressive, with a booming voice, as I remember. Not very many people would take issue with his thoughts or expressions, even if they disagreed violently, because he was kind of an overwhelming personality.

Scott: He was intimidating?

Crandall: Yes. That is one word for it. I think so. I did not see anyone stand up and disagree with him. But he was one of the principals in early knowledge of earthquake and seismic considerations for structures. I

guess those who had classes from him kind of revered the guy.

Hardy Cross

Crandall: You mention Moore's reference to moment distribution, which he probably learned as a student of Martel's at Caltech. That brings up the name of another great engineering professor, Hardy Cross.[14] I remember the Hardy Cross moment distribution method that I was taught while a student at Berkeley. You'll recall I graduated in 1941, so this was way before the modern computers engineers now use. Hardy Cross had developed an approximate method that made it possible to solve some complex problems with simplifying assumptions. Say you had all these structural forces coming in on a girder and column. You put weight on a continuous framing, frames with relatively rigid column-beam joints, and the bending moments flow all over. What is the distribution of the moments and forces? It was a perplexing problem to solve with a slide rule unless you had an elegant conceptual approach, which Cross developed.

It was a very clever method. I didn't work with it that much, since it was mostly for structurals. It was also applicable to hydraulics, where

---

14. Hardy Cross (1885-1959), obtained his master of civil engineering degree at Harvard in 1911. After being a professor at Brown University and briefly a consulting engineer, he joined the faculty of the University of Illinois at Urbana-Champaign in 1921, where he developed what became known as the Hardy Cross Moment Distribution Method ("An Analysis of Continuous Frames by Distributing Fixed-End Moments," *Proceedings of the American Society of Civil Engineers*, May, 1930).

you had multiple pipes coming together and you needed to figure how you distributed the flow from one pipe to the other pipes.

Scott: Evidently it was used considerably, probably from the early 1930s up to when the computers came in.

Crandall: I don't know whether something else replaced it prior to the age of the computer, but it seemed to me that it was the only way structural engineers could get a reasonable estimate of the distribution of forces in a frame structure.

Robert V. "Cap" Labarre

Scott: What about the [Cap] Labarre and [Fred] Converse consulting firm?

Crandall: I never got to meet Labarre. He passed on right after I came to Dames and Moore. He apparently was a legend. His name was Robert, but everyone called him "Cap" Labarre. He apparently was the first guy to actually practice soil engineering and foundation engineering in southern California. I think he came from Louisiana.

The Field Act came along, and Labarre got involved in the school program. He saw an opportunity, say about 1935. He would come up with reports for the foundation design. Essentially, the report consisted of the allowable bearing value for the soil supporting the foundation.

He did it by the load test method. This was an early technique. Much of the time you took a 12"x12"-square post, set it on the ground vertically and started loading it. When the post started punching in the ground, that was the

ultimate bearing value of that soil. You divided that by two or some such safety factor number, and wrote a report saying, "This soil is good for 1,250 pounds per square foot," or whatever it was. That was the way they did it originally. Then I think Labarre got started with the exploratory boring work. Both Trent and Bill worked for him, I think, while they were in school. They got interested in soil engineering and got part-time or summer jobs with Cap Labarre and then graduated. I think Dames worked for Labarre, and Bill Moore went to work for the Corps of Engineers. Then Trent started his own company, and as soon as he got some work, Bill left the Corps of Engineers and it became Dames and Moore. I think that was in 1938 or 1939.

Karl Terzaghi: Father of Soil Mechanics

Crandall: I have to digress to discuss Karl Terzaghi before going on to talk about Fred Converse, because Fred got interested in soil engineering and went back to Harvard when Karl Terzaghi[15]—the father of soil mechan-

---

15. Karl von Terzaghi, (1883-1963) grew up in Prague and went to college at the Technische Hochshcule (technical university) in Grasz, Austria, receiving a degree in mechanical engineering. His book, *Erdbaumechanik*, or *Soil Mechanics* in English, was published in 1925 based on research conducted as a professor in Istanbul. In 1925 he was hired by MIT, then moved back to Austria, then immigrated again to the USA just prior to the outbreak of World War II, where he was on the faculty of Harvard University till his retirement. He also consulted on a number of large dams and other projects.

ics—had come to the United States. Terzaghi had come up with his theories of consolidation and other things, and was really the first one to put soil engineering on a scientific basis. Up to that time, if an adjacent building didn't fall down, you did what they did, or maybe you added a little greater soil pressure, until something happened, and then you backed off.

Terzaghi had come up with this theory of consolidation and some sensible approaches to soil engineering. He gave the soil engineer the analytical tools to understand how soil behaved.

He had a few disciples, one of whom was named Arthur Casagrande. Casagrande took over at Harvard when Terzaghi left the scene.

Going back to Labarre and Converse, Fred Converse was a civil engineering professor at Caltech. He had gone to Harvard and absorbed the influence of Terzaghi and Casagrande there. Then Converse started teaching a course in soil mechanics at Caltech. I think Trent Dames and Bill Moore took that course when they got their masters degrees there.

Fred Converse was doing a little consulting, and he and Labarre joined forces. They formed Labarre and Converse, which gave a little more scientific credibility to the work that Labarre had been doing. At this time, they started taking so-called undisturbed samples. They drove a cylinder—something like a pipe, about two-and-one-half inches in diameter—into the ground and then extracted it, and ran laboratory tests on the sample. This is what we still do today, although there has been some improvement in the sampling design and procedures.

Cyclic loading was done with a static load test, which did not have the capability of very quick on and off loadings or vibration tests. You did the static test with a dead weight frame supporting a mass of concrete or steel or something—to give the resistance, the reaction—and a hydraulic jack pushing against that, and the other end is pushing on this bearing plate that is on the soil. You could cycle the loadings as fast as you could pump the jack up and let it off. It was maybe a minute between cycles, or maybe two minutes, nothing like the split-second loading reversals we can do now.

Warren and Converse

Crandall: Those were the early years. Fred Converse kept his job teaching at Caltech. After Labarre either died or retired, the firm became Warren and Converse for a while. Donald R. Warren I think at one time was State Highway Engineer, and had done a lot of bridges and things for the state. He was doing a lot of designing in Los Angeles, and added the soil engineering capability by joining forces with Converse.

When I joined Dames and Moore, our competition was Warren and Converse. Warren and Converse came to some disagreements and split up, and Warren kept on in the soil business, as well as his structural design business. That firm was one of the ones hit hard by lawsuits from work on those housing tracts we discussed earlier.

Long-Term Liability Exposure

Crandall: Warren and Converse did most of the tract work in Los Angeles at about that time, in 1941, when I started with Dames and Moore and in the following years. Believe it or not, as recently as the 1980s, when the company was still active, they were still getting lawsuits from that work. A landslide would cause damage and the cry would go out: "Go get the soil engineer."

Scott: That was a full forty years later!

Crandall: Oh, yes. Projects can come back to haunt the soils engineer decades later. Of course the developer is gone, or that corporation has changed. The designer of the building, the architect, is probably gone. The earthmover is out of business. The only guy left is the soils engineer, and these poor guys are getting lawsuits. Well, it has happened to Dames and Moore, and to me, for work done way back. Now of course they judge it by the current code. You think that the statute of limitations should protect you, but that does not do you any good until the problem occurs, and then they start measuring from that time.

Scott: They start the clock with the occurrence of the problem, not from the time the work was done?

Crandall: Yes. And usually what happens is that the homeowner or property owner changed something, or did not take care of the drainage system or something like that, and is looking for somebody to help pay for the costs.

Scott: But even if the owner is responsible, you have to fight it through in court to demonstrate his responsibility?

Crandall: Right. I felt sorry for Warren and Converse because in a tract of maybe one hundred homes, ninety-nine might be perfect, but one has a problem, and maybe that problem is not even of your doing. But to the one guy who owns the home, it is the biggest thing in the world, being his main investment. When something goes wrong, the owner gets the lawyers to put the noose around anyone they can find, and with the developer long gone, often the soil engineer is who they find. We're in a terrible business from that standpoint.

Chapter 10

California Seismic Safety Commission

I consider the work of the Seismic Safety Commission an extremely valuable and important effort on behalf of the people of the State of California.

Scott: We've both served on the California Seismic Safety Commission. Say a few words about your experience on it.

Crandall: Well, Stan, it's you who should be talking, since you were on it from the very beginning. I consider the work of the Seismic Safety Commission an extremely valuable and important effort on behalf of the people of the State of California. The Commission has taken the lead in emphasizing the importance of earthquake preparedness, proper design, and effective building codes and their enforcement.

Most of all—I think any good practicing engineer would tell you this—is the extreme importance of inspection during construction. Beautiful plans, computer printouts, and state-of-the-art seismic knowledge will not protect your structure if it is not built in accordance with those plans and computations. Sometimes it only takes one weak point to cause a catastrophe that otherwise would only have been a minor incident in the life of a building. The Commission has had some success in increasing quality control in construction also.

Mines and Geology Board, the Alquist-Priolo Act

Crandall: Preceding my relationship to the Seismic Safety Commission, my first assignment in state government as an appointee of any significance was to the state Mines and Geology Board. That appointment was made under Governor Ronald Reagan in 1973. I served there at a crucial time with some very fine and talented people—like Dick Jahns and Clarence Allen, two outstanding geologists. I served as a soil engineer. The Mines and Geology Board is part of the California Division of Mines and Geology. One of its roles was to provide advice on carrying out the Alquist-Priolo Special Studies Zones Act of 1972.[16]

The Alquist-Priolo bill—pushed by Alfred Alquist (whom I mentioned earlier) in the California Senate, and Paul Priolo, in the other legislative branch, the California Assembly—came about in 1972. It was passed in order to identify active faults in the State of California that posed a surface rupture hazard, to make the public aware of their location, and to require local governments prior to issuing building permits to have geologists prepare studies of sites located in zones mapped by the state where the hazard might exist. A fault zone

designation in an area does not mean that you cannot build on that area, but it means that a geologist registered in California must study the problem. If your proposed development falls within the zone mapped by the state, your local building agency is required to receive this study, paid for by the owner, to determine whether the development is safe.

Believe it or not, one of the things that triggered that bill was the knowledge that subdivisions in San Bernardino were being built right on the San Andreas fault trace. That is sort of astounding, because one of the things you learn in engineering is to avoid building across a fault. The dramatic surface faulting in the 1971 San Fernando earthquake brought attention to this kind of hazard.

Theoretically, you can design a structure to resist most any level of ground shaking from an earthquake, but there is very little you can do to resist rupture of the ground beneath the building. Most geologists, from a theoretical science point of view, would like to identify any crack in the earth as a potentially active fault. Geologists think in terms of geological time—millions of years—and all kinds of things can happen in such long time spans. If you look hard enough, you can find faults almost anywhere in California. By digging down, looking at oil well maps and logs, you can find faults of all sorts, most of which don't get to the surface, and hopefully never will. However, that doesn't give you a practical basis for establishing hazard zones for surface rupture. Clarence Allen, Dick Jahns, and I did much screaming and hollering, and were able to limit the delineation of the faults to the ones that were active in Holocene time—that is, they

---

16.   The law was later re-named the Alquist-Priolo Earthquake Fault Zoning Act.

had displaced during that time. The Holocene epoch is roughly the last 11,000 years. That policy defined an active fault for purposes of the Alquist-Priolo Act. That was for ordinary building developments. For critical facilities like dams, there is a rationale for extending that definition to faults whose most recent rupture may have been a lot farther back.

Appointment to the Seismic Safety Commission

Crandall: I was reappointed for another term on the Mines and Geology Board by Governor Ronald Reagan. When Harry Seed resigned from the seat on the California Seismic Safety Commission that was designated for soil and foundation engineering, now called geotechnical engineering, I was appointed. The seat is for a civil engineer specializing in soils and foundations. The Commission also has seats set aside for a geologist and a seismologist along with slots for local government, emergency services, and so on.

Scott: Harry Seed had occupied that post from the time of the original formation of the Commission, when it was set up by the Seismic Safety Act of 1975, up until he either resigned from the Commission or declined to be reappointed.

Crandall: Yes. I was appointed in 1982, when Edmund G. Brown, Jr. (Jerry) was Governor of California. That was when you were chair of the Commission, Stan.

In addition to attending Commission meetings, which I have done pretty regularly, my main participation in connection with the

Commission has been through the Strong Motion Instrumentation Advisory Committee, which we discussed earlier. I took over as chair after Bruce Bolt in 1984.

Scott: For the record, in my view, of all the Commission's committees, that is by far the biggest operation, and is a continuing operation. All the other committees are set up and operate for one or two years, maybe for five years or so, and then go out of existence.

Crandall: Yes, the other committees have an assignment to write a report, or something like that. You are right about the scope and duration of the strong motion committee work. The budget of the strong motion program is many times what the Commission's budget is.

Observations on the Commission

Scott: You have provided a lot of background on the strong motion program. But tell me about your general observations and comments on the Commission itself.

Crandall: There have been three executive directors of the Commission that I have known. Bob Olson was the first, then Dick Andrews, and now as of 1991, Tom Tobin. I had some experience under all three. I think Bob was just leaving or had just left when I came on board, so it would was mostly Dick Andrews, and Tom Tobin.

Scott: I think Bob Olson left in 1982.

Crandall: I remember I was appointed at the same time that Bill Iwan was, as well as the lovely lady who had a position with the Red Cross and later moved to Washington, Ann

Boren. The three of us met at my office, at the request of Bob Olson and Dick Andrews. They gave us a little background and kind of a briefing session, and so Bob Olson participated in that. Perhaps he had recently left the executive director job to launch his consulting career, but he came to the meeting with Dick Andrews and the two were kind of in a transitional period. Tom Tobin became executive director a little later.

I can say this without any reservation: I think they would have to search long and far to find someone that is better suited for this job than Tom Tobin is, believe me. I have great admiration for what he has done. From my viewpoint everything I have seen of Tom's work has just been most commendable.

The briefing session participants were the new Commissioners, Bill Iwan, Ann Boren, and I, plus Bob Olson and Richard Andrews. We sat in the conference room in my office. Bob and Dick did the talking. Mostly it was Bob, explaining a little of the history of the Commission and its purpose, what it was doing, what it was trying to do, and what would be expected of us as Commissioners. He did a good job of clueing us in on the nature of the beast.

I don't know if the briefings are still being done at this time, but it was a worthwhile thing to have that kind of an introduction before we actually got involved in Commission business.

Scott: I would think briefings would be quite valuable, but I really don't know exactly what is done now.

Crandall: This did not happen when you were a new commissioner? Well, you were one of the first Commissioners, in at the inception of it.

Scott: Yes. I was on at the outset, and there was no briefing that I recall, beyond a meeting with the Governor, and then we met together as a group for the first time and talked about what we ought to do.

Crandall: Regarding the quality of the Commission, it is amazing, the type of people that have been appointed to this Commission, who in my opinion are outstanding. Both as citizens and as professionals. They were able to get really top talent to serve in this capacity. You, for example, who have served from the beginning—the only one still serving, I believe.

Scott: Of the original group of Commissioners, I am the only one left, although Bruce Bolt has also been there a long time. He must have come along three or four years after the Commission started up. So Bruce has been on twelve or fifteen years, and there may be another one who has been around a long time.

Crandall: It is a great public service, I think. That is why I am interested in it—I believe that the things they do are just incredibly important. I am amazed at the amount of output, and the quality of the reports and the work that's done. I am not good at the workshops, although I realize that those are very important things. While I have attended every one, I think my contributions have been really limited.

Scott: You are talking about the annual two-day or day-and-a-half workshop?

Crandall: Yes, where we brainstorm, or discuss what we should be doing. I don't seem to be creative in that kind of thinking. Maybe I represent the ordinary citizen in that regard, who needs to be shown, and once you see what the path is, maybe you are able to provide a little light along the way, as far as picking out the path.

Scott: Are you saying that you play more of a listening or reacting role in the workshops?

Crandall: I guess, in a sense, yes.

Scott: I don't have that impression. At least I don't think of you as playing a shrinking violet role.

Crandall: Well, if I get stepped on, I holler.

Scott: As to your comments about the quality of the Commissioners, can I ask you a leading question? It has been my impression, and I have gotten similar feedback from some other Commission members, that we did not universally get the best talent. I have heard complaints, particularly during the Jerry Brown administration, that some appointments were not necessarily the best the state could have gotten. Do you have that feeling?

Crandall: My comment and knowledge has been limited to the technical type personnel, like Bruce Bolt, Lloyd Cluff, Bill Kockleman, Bill Iwan. All of those are outstanding technical people. Perhaps I should have qualified my statement, because I am not that familiar with the other type of appointees. I know your work, and I cannot think of anyone that I have higher regard for in the work of the Commission. By the way, what slot were you appointed to fill?

Scott: Surprisingly enough, I am representing local government, at least theoretically. There is an interesting story on that which I'll sum up here. After the first round of appoint-

ments that established the initial membership of the Commission, the regular process then went into effect, defining designated seats for people from various backgrounds, and there was no slot for me.

Of the very first group of appointments, half were basically chosen by Senator Alquist on behalf of the Joint Committee—but really they were selected by Karl Steinbrugge and Senator Alquist followed his suggestions. The other half were chosen on behalf of the Governor by Jim Steams, the Governor's Secretary of Conservation. They did not have to fit precisely in the disciplinary appointment slots that have applied since those first Commissioners were chosen in late 1974 or early 1975. Anyway, half the first Commissioners were appointed on Senator's Alquist's side, with Karl Steinbrugge calling the shots, and the other half were selected by the administration's side. I was one of the Steinbrugge/Alquist appointments.

That was how the initial set of appointments was made. But later, when reappointments or replacements came up, the bill's regular legal formula took over. There being no seat with qualifications I possessed, Bob Olson and the Commission chair, probably either Karl Steinbrugge or Bob Rigney at that time, arranged for the League of California Cities to recommend my appointment as a city government representative.

Crandall: I have not really evaluated the other members of the Commission. I know that I have great respect for several, technically. Those I have known or know well in the field include Bruce Bolt, Al Blaylock, who was a member and then resigned to become a member of the state licensing board for engineers. Also Paul Fratessa—he was a strong member. There was Bill Waste, from insurance—I did not know him before, but Bill Waste was an impressive guy.

I think it is a good, well-balanced Seismic Safety Commission. You need viewpoints and various perspectives, of course. I personally think in a technical way most of the time, but there are other aspects of problems. Some things I did not realize until I got involved in the Commission—including the social impact, such as with the unreinforced masonry buildings and their retrofits. As an engineer I would say, "Hey, those old brick buildings should come down—demolish the unreinforced masonry structures, get them out of here." But then, as we learned by visits to Chinatown in San Francisco and in other ways, those buildings house low-income people, and what do you with those people? Maybe you make it worse for them. One of the interviewees told us, "These dispossessed people are going to die from pneumonia on the streets faster than they will from an earthquake collapse." So you get a broader view. You see that you cannot tear things down just because they are less safe than something else, without having impacts in other areas. I think I have broadened my understanding and tempered my positions.

In my years as an engineer, most of the time I thought, "Hey, engineers can solve all these problems." Let us build buildings that will not collapse, and get rid of the old unreinforced masonry buildings that will collapse. Then suddenly you start thinking, "Maybe we cannot

do that overnight. There are other concerns that are also important."

Scott: Do you have other impressions of the Commission or its activities, or things that you have been involved in or that you have watched, or that you think the Commission ought to be involved in?

"California at Risk": The Ripple is Spreading

Crandall: You bring up things I have never thought about. On the other hand, I think the Commission is doing many things. For example, take the "California at Risk" program. A tremendous amount of thinking is going into that.

Scott: Yes, and it is activating all kinds of groups and agencies out there, maybe slowly and gradually in some cases, but nevertheless activating them.

Crandall: The ripple is really spreading now, believe me. I am talking now about things I am reasonably familiar with, and this could be happening in other committees and disciplines as well. For example, the peer review process was part of what was decided would be a good thing in the study of the state's prison construction program. But the peer review process not only was adopted by the prison committee, but has also found its way into other areas of the State Architect's office, and the University system has realized the importance of such review.

One thing that impresses me is the high regard that the Commission must be given, the high esteem in which the Commission is held by the legislature and others in the state government. To be an advisory group to these various state operations is a very noteworthy recognition of the Commission.

With respect to the purpose of the Commission and the talents of its membership, I am completely impressed by the people who are serving as Commissioners, and by the competence and abilities of these people, and their willingness to cooperate.

Scott: Why don't we talk more about the appointment process for choosing Commission members? What do you know about how your own appointment was made? I think in some cases some potential members actively sought membership on the Commission, whereas in other cases they probably did not advance their candidacies at all, but other people were active on their behalf. There are probably all kinds of combinations and variations.

Crandall: I am pretty naive about all of this. I'll be darned if I know for sure how that happened in my case. The Commission's executive director probably seeks advice from organizations that are representative of the position in question, and those organizations put forward a name or two as possible appointees. This probably happened in my case, but I don't remember being aware that I was under consideration for the assignment.

Before I was appointed, of course, I was asked if I would accept, and I think I might have answered a questionnaire and was informed about the ethics and the conflict-of-interest matters that are important in such a public position. But my guess is that Bob Olson or possibly Dick Andrews did a little solicitation

for suggestions, probably from the American Society of Civil Engineers. The structural engineers may also have been asked, and the Consulting Engineers Association of California, and also the Geotechnical Engineers Association. My name probably got proposed that way, but I am not sure. Maybe somebody already on the Commission was aware of my background to the point that my name was submitted.

Scott: The suggestions were probably then sent to the Governor's appointment secretary. That was during the Jerry Brown administration, was it not? Jerry Brown had the reputation of being slow on appointments.

Crandall: I think he did not believe in commissions, and as a result did not pay attention. That was a terrible thing for the state government, I think. Many of the boards and commissions could not function without a quorum, and if he did not appoint somebody, maybe a quorum did not exist.

In any event, I guess I was appointed first by Governor Jerry Brown and reappointed by Governor Reagan. I don't know exactly how it worked, but of course before you are actually appointed you must submit financial disclosure information, and be aware of conflict-of-interest regulations and matters of that sort. I was never questioned about my party affiliation or political views.

Scott: You can say in Jerry Brown's favor that when he finally got around to appointing somebody, apparently he did not let the decision be influenced a lot by political registration, at least in my case.

Crandall: Somebody did a good job in getting a cross-section of both technical and social concerns on the Commission. It is very well-balanced. Barbara Riordan, the current chairperson, for example, is a county government representative, and you represent city government.

Chapter 11

Concluding Words in 2000

...while the computer can do many things, it must have reliable data with which to work. Such data are now available from strong motion records...

Scott: It's been a few years since we recorded our interviews back in 1991. This is a chance to bring the story up to date, as of early 2000. What would you like to add to complete the interview series?

Crandall: First, let me say I greatly appreciate the time and effort that you have put into this, Stan. I have seen the final product for several of your other oral history publications in the EERI *Connections* series, and they are indeed superb. I just hope that this one will develop at least part of the interest that the others have generated.

I think it would be worth reviewing some of the important seismic events that have occurred since we last spoke in 1991. Probably the main event was the Northridge, California earthquake in January of 1994. I believe that more useful records were obtained of ground motion and building behavior from that earthquake than ever before.

The instruments operated by SMIP—the state's Strong Motion Instrumentation Program—and also those installed by other agencies, provided the kind of information for which structural engineers have long yearned. I feel more than ever that the strong motion program will continue to provide great benefits to the public in the better understanding of the performance of our buildings under earthquake conditions. Not only will the buildings be safer, but also we will be able to design for the actual forces in a more economical manner. Those results have been very gratifying to those of us who worked for many years to see the state establish a comprehensive Strong Motion Instrumentation Program. I have not, however, been directly involved in the SMIP Northridge earthquake data utilization, as my membership on the SMIP Advisory Committee terminated shortly before the Northridge earthquake, when my service on the Seismic Safety Commission ended in 1993.

Scott: You had been associated with the SMIP program since 1972, and you also served on the Commission a total of 11 years. Those pioneering efforts have certainly paid off in useful results that have served the profession very well. What other comments would you like to make about some of the main seismic developments since our last interview?

Crandall: You and I talked a decade ago about how the advent of the computer has provided design engineers with a very strong tool, an opinion that has been more than justified by subsequent developments. Computer usage and programs have far exceeded what appeared to be on the horizon back at the time of our last interview. Again, however, while the computer can do many things, it must have reliable data with

which to work. Such data are now available from strong motion records that show the detailed behavior patterns of various types of structures. Given the structural behavior as recorded by the strong motion instruments, the computer can now be used to perform incredibly detailed analyses of such behavior. This is definitely leading to much better design, and improved construction techniques for future building performance under earthquake conditions.

In just my own field of geotechnical engineering, the potential for predicting the ground motion conditions at individual sites has been greatly enhanced. The anticipated earthquake in the Parkfield area has failed to occur, preventing us from obtaining results of the several experiments programmed for that expected earthquake. With luck, valuable information may still be developed when and if the Parkfield event does take place. I realize it may sound a little strange to appear to be hoping for an earthquake to occur. We had great expectations, however, for the kinds of information that the various Parkfield test installations will provide whenever the next earthquake does occur. Nevertheless, as I suggested above, the data we did get from actual structures instrumented in the SMIP have been very beneficial.

Scott: Bob Wallace discussed the Parkfield earthquake prediction experiment in his EERI oral history, published in late 1999.[17] According to Bob, much has been learned from that experimental effort, despite the earthquake's

17. *Connections: The EERI Oral History Series—Robert E. Wallace*, Stanley Scott, interviewer. Earthquake Engineering Research Institute, Oakland, CA, 1999.

failure to happen. But of course, the most important kinds of things structural and geotechnical engineers hoped to learn about do depend on earthquake shaking, which has not occurred.

Since we last spoke, a number of major earthquakes have occurred in different areas, among the recent ones being the 1999 earthquakes in Turkey and in Taiwan. Do you have any comments on those?

Crandall: I have not been too closely involved in those major recent earthquakes. They certainly were devastating in the areas where they occurred. From what I have read in reports by the investigators from the United States who visited those sites, much of the destruction was due to the nature of the structures involved. That kind of structural destruction is similar to what we have observed in other countries where the construction methods used, especially in the older buildings, do not provide much earthquake resistance. Still, the teams that visited these sites learned useful information that will be applicable to some of the conditions in our own area. Also, what has been learned from the liquefaction and ground displacement that occurred during those earthquakes is applicable to all areas having similar characteristics. As a result, the geotechnical investigators were able to obtain important information that will be directly applicable to sites in California.

Scott: Talk a little about what has happened to you personally in the years since we last talked.

Crandall: When we last spoke in 1991 I was an employee of Law/Crandall. You will recall that the company increased the scope of its services to include environmental and construction materials services, among other features. As a result, it seemed desirable to change the original company name from LeRoy Crandall and Associates to Law/Crandall. This took place in 1991.

I have functioned independently of the main company since 1987. My direct office was concerned entirely with forensic matters, specializing in construction defect litigation. On January 1, 1999, I retired from Law/Crandall and formed a new company called Crandall Consultants, Inc. Under that designation, I am still operating as a forensic consultant in geotechnical engineering. My activities are almost completely concerned with litigation, with very little involvement in the design aspects of geotechnical engineering.

My one regret is that I have been so tied up in the business activities that I have been unable to attend and be active in the professional societies to the degree that I have been in the past. I still maintain membership in all of them, but it is difficult to find time to attend the meetings and conventions. The principal drawback in the litigation field is one's inability to maintain a definite schedule. If a case goes to trial, the experts are expected to be available to meet the court requirements. Also, arbitration and mediation sessions are obligatory, and are often scheduled at the last minute. As a result, it is hard to plan for attending professional meetings and engaging in related activities.

I still find the work very interesting, so much so that I still put in at least forty hours a week. My one concession to getting older is that I now try to avoid working on Saturdays and Sundays.

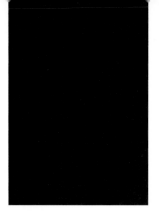

Photographs

Graduation from San Diego High School, California, 1935.

LeRoy Crandall (left) and his brother Clifford with their grandfather, Jefferson L. Crandall, 1922.

LeRoy Crandall, with fiancée Eileen Exnicios, at Crandall's graduation from the University of California at Berkeley, 1941.

Crandall at the office of LeRoy Crandall and Associates at 1619 Beverly Boulevard in Los Angeles, California, 1955.

Eileen and LeRoy attend a costume party for the Structural Engineers Association of California, 1959.

Disneyland was one of the first geotechnical engineering projects the new firm of LeRoy Crandall and Associates undertook, 1954.

Partial excavation for a high-rise being built in downtown Los Angeles, California. LeRoy Crandall and Associates did the geotechnical engineering, 1965.

The aerial tramway at Palm Springs, for which LeRoy Crandall and Associates provided the geotechnical engineering, 1961.

The tie-back shoring system (partial depth) for the Century City Theme Towers, which was the deepest excavation ever attempted at that time, 1970.

The San Bernardino County Foothill Communities Law and Justice Center, Rancho Cucamonga. This was the first building in the United States to use base isolation. LeRoy Crandall and Associates provided the ground motion design criteria, 1987.

Elected to honorary membership in the American Society of Civil Engineers, 1984.

LeRoy Crandall and Associates moved their office to Glendale, California in 1986.

Crandall in his office at 711 North Alvarado Street in Los Angeles. LeRoy Crandall and Associates commissioned the building and maintained their headquarters office there from 1966-1986 (photo circa 1984).

Aerial view of downtown Los Angeles, California. LeRoy Crandall and Associates was the geotechnical engineering firm for all but two of the high-rise buildings (photo by Marshall Lew).

Above: *Eileen and LeRoy at the Taj Mahal, 1993.*

Left: *Eileen and LeRoy on a cruise in 2000.*

Marshall Lew and LeRoy Crandall, January 2008.

A

Accelerographs, 49–50

Airfields and airport facilities, 16–17

Allen, Clarence R., 81, 94

Alquist, Alfred E., 53, 58, 94–95, 97

Alquist-Priolo Special Studies Zones Act of 1972, 94–95

American Society of Civil Engineers (ASCE), 53, 79, 99

award from, 7

Crandall member of board of directors of, 79

Andrews, Richard, 95, 99

Asphaltic sands, La Brea Tar Pit, Los Angeles, California, 46

Attenuation of strong ground motion, 61

B

Banville, Joe, 11

Barnes, Fred, 23, 26, 29, 30

Barnes, Steve, 85, 87

Base isolation, 68–70

bridges, 69

San Bernardino County Foothill Communities Law and Justice Center, Rancho Cucamonga, California, 68, 109

University of Southern California Hospital, Los Angeles, California, 69

Bay Area Rapid Transit (BART), 81

Beanfield, Rufus M., 87

Bearing ratio tests of airfield pavings, 16

Beerman, Paul, 7

Bentonite clay, 35, 44

Blaylock, Al, 97

Blume, John A., 49, 50, 53, 86

Bolt, Bruce A., 57–58, 95, 96, 97

Boren, Ann, 96

Bowen, Oliver G., 85, 87

Brandow and Johnston, engineers, 87

Brewer, William (Bill), 7, 9, 12, 16

Bridges

base isolation of, 69

Caltrans, 59–60

Golden Gate Bridge, 59

lifelines, 81

San Francisco-Oakland Bay Bridge, 59

Vincent Thomas Bridge, Los Angeles, California, 59

Brown, Jr., Edmund G. (Jerry), xiv, 95, 99

Brown, Glenn A., xiii, 30

Brown, Glenn A. and Associates, merger with Crandall firm, 30

Brugger, Walter, 55

Buildings, modeling of, 85

Butts, Wendell, 87

C

C.F. Braun & Co., 20

California at Risk program, 98

California Division of Mines and Geology (CDMG), 54, 71, 82, 94–95

Crandall board member of, 95

California Office of Statewide Health Planning and Development, 61

California Seismic Safety Commission (CSSC), 54, 60, 68, 86, 93–99

California at Risk program, 98

Crandall member of, 56–59, 93–99, 102